Cover illustration: photograph of a printed circuit board cross section, courtesy of Young H. Kwark, IBM T. J. Watson Research Center, Yorktown Heights, New York, USA

Bibliografische Information der Deutschen Nationalbibliothek

Die Deutsche Nationalbibliothek verzeichnet diese Publikation in der Deutschen Nationalbibliografie; detaillierte bibliografische Daten sind im Internet über http://dnb.d-nb.de abrufbar.

ISBN 978-3-8325-2776-1

Logos Verlag Berlin GmbH
Comeniushof, Gubener Str. 47,
10243 Berlin
Tel.: +49 (0)30 42 85 10 90
Fax: +49 (0)30 42 85 10 92
INTERNET: http://www.logos-verlag.de

Development, Validation, and Application of Semi-Analytical Interconnect Models for Efficient Simulation of Multilayer Substrates

Vom Promotionsausschuss der
Technischen Universität Hamburg-Harburg
zur Erlangung des akademischen Grades
Doktor-Ingenieur (Dr.-Ing.)
genehmigte Dissertation

von

Renato Rimolo-Donadio

aus

San José, Costa Rica

2010

1. Gutachter:

Prof. Dr. sc. techn. Christian Schuster

2. Gutachter:

Prof. Dr.-Ing. Arne Jacob

3. zusätzlicher Gutachter:

Dr. Xiaoxiong Gu (IBM T. J. Watson Research Center, NY, USA)

Vorsitzender des Promotionsverfahrens:

Prof. Dr. Ernst Brinkmeyer

Tag der mündlichen Prüfung: 17.12.2010

About the illustration:

Urban Interconnects

by R. Rimolo-Donadio,

Acrylic on wood, 2002.

Abstract

This thesis deals with the efficient modeling and simulation of multilayer substrates in high-speed electronic systems, such as packages and printed circuit boards.

Semi-analytical models for the electrical behavior of vias and traces are presented and a framework for automated simulation of multilayer structures is proposed. The models are devised in terms of microwave network parameters and they rely on the formulation of the parallel-plate impedance to describe wave propagation between adjacent reference planes. Via-to-plane capacitances are used to approximate the near fields around via barrels. A modal decomposition method allows the merging of parallel-plate and trace models; microwave segmentation techniques are applied to solve multilayer configurations.

An extensive and thorough validation of the models is presented, using general-purpose numerical methods for electromagnetic simulation and hardware measurements. The validation cases include multilayer via configurations with power and ground vias, mixed reference planes, single-ended links, differential links, and via arrays. The numerical efficiency, advantages and disadvantages of the proposed approach are covered in the discussion.

Several application scenarios of realistic complexity are also evaluated. Studies of differential links between ball grid arrays, stub via resonances, and differential to common-mode conversion are presented. The utilization of the models for co-simulation of power and signal integrity is demonstrated as well as the extension of the method to handle arbitrarily shaped plates and radiated emissions. It is shown that the models can provide good results up to 40 GHz and a numerical efficiency of at least two orders of magnitude better than general-purpose numerical methods for electromagnetic simulation.

Author key-words: modal decomposition, printed circuit board, package, parallel plates, power integrity, signal integrity, traces, via.

Acknowledgment

This thesis was the result of three and a half years of work at the Institute of Electromagnetic Theory (TET) of the Technical University of Hamburg-Harburg (TUHH), between November 2006 and May 2010, with a research position funded by TUHH. Along this time several people and organizations have contributed in different ways to the completion of this work.

I would like to express my sincere gratitude to Prof. Dr. Christian Schuster, for giving me the opportunity of carrying out this work, and for the dedicated guidance and advice. I was lucky of having an extraordinary person and scientist as supervisor. His motivation, high principles, and commitment to the research and academic activities have been the model to follow during all this time. I would also like to thank Prof. Dr. Arne Jacob, second examiner, and to Prof. Dr. Ernst Brinkmeyer, president of the Doctoral Committee evaluating this thesis, for the careful revision and evaluation of this work.

My gratitude goes to Dr. Xiaoxiong (Kevin) Gu, who was also external examiner of this work. His support and feedback along the different phases of this project have been essential for its successful completion. I would also like to thank Dr. Young H. Kwark, for his mentorship, support, and all the valuable feedback provided during these years. My gratitude is extensive to Dr. Mark B. Ritter, M. S. Christian Baks, and the other members of the former High-Speed I/O Subsystems and Packaging Group at the IBM T. J. Watson Research Center, Yorktown Heights, New York, USA, who have cooperated a lot with our research and provided the hardware and measurements used in this work. I also had a great and productive stay with them at IBM during the summer 2007.

This work would not be possible without the help of Dr.-Ing. Heinz-Dietrich Brüns. His contributions to the research activities, careful analysis of the work, and feedback have been crucial for the development of this project. I would also like to express my gratitude to all the colleagues and staff at the Institute of Electromagnetic Theory for their support. Special thanks to my colleagues Dipl.-Ing. Miroslav Kotzev, M.Sc. Xiaomin Duan, and Dipl.-Ing. Sebastian Müller, for their help and the great team work. Thanks go also to the students who contributed with their final works to our research activities.

The discussions with the people involved in the IBM/ Missouri University of Science and Technology (MST)/ University of L'Aquila/ TUHH weekly meetings have been also very important for the progress of this work. My appreciation goes to all them, in particular to Dr. Bruce Archambeault, at IBM USA, Prof. Dr. James L. Drewniak, Prof. Dr. Jun Fan, Dr. Yao-Jiang Zhang, Prof. Dr. Albert Ruehli, and their research team at the Electromagnetic Compatibility Group, MST (former University of Missouri-Rolla), USA, for all their cooperation and feedback.

Finally, I would like to thank to my family and friends. I am indebted to my wife Karolina and our son Leonardo, my parents Roque and Olga, and my sister Fiorella for their unconditional support. This thesis is for them.

Contents

List of Figures and Tables

Figures

Tables

List of Symbols and Acronyms

Notation

V	Scalar
$\overline{V}$	Matrix vector
$\overline{\overline{V}}$	Matrix
V_{ij}^{pp}	Matrix entry, where indexes i and j denotes the matrix row and column respectively, and the superscript the type of variable
x, y, z	Cartesian coordinates
ρ, φ, z	Cylindrical coordinates
u	Superscript for upper/top side of a cavity
l	Superscript for lower/bottom side of a cavity
tl	Superscript for transmission line modes
pp	Superscript for parallel-plate modes
$\boldsymbol{R}$	Spatial vector

Symbols

ω	Angular frequency
α	Attenuation constant
T_b	Bit time
C	Capacitance
d	Cavity thickness
Z_0	Characteristic impedance
G	Conductance
σ	Conductivity
t_p	Conductor thickness (for reference planes)

t_c	Conductor thickness (for signal layers)
I	Current
$i\,(t)$	Current (time domain)
f_c	Cutoff/ transition frequency
$\tan\delta$	Dielectric tangent loss
dg	Distance between a signal and a ground via (center-to-center)
ρ_{ij}	Distance between ports (radial waveguide method)
h^l	Distance from trace to bottom reference plane
h^u	Distance from trace to top reference plane
$\mathfrak{E}$	Electric field
f	Frequency
f_s	Fundamental frequency
$H_n^{(1)}$	Hankel function of first kind and order n
$H_n^{(2)}$	Hankel function of second kind order n
h	h-parameter
$\overline{\overline{E}}$	Identity matrix
j	Imaginary unit
u_{ind}	Induced voltage
L	Inductance
L_{interc}	Interconnect inductance (for decoupling capacitors)
a, b	Lateral dimensions of rectangular plates (in x and y, respectively)
$\mathcal{H}$	Magnetic field
f_{max}	Maximum frequency of interest
f_{min}	Minimum converged frequency
T_{md}	Modal decomposition matrix
k	Modal decomposition transformation factor
Y^{pp}	Parallel-plate admittance
Z^{pp}	Parallel-plate impedance
T	Period
μ	Permeability (when subscripted, d: dielectric, c: conductor)

μ_0	Permeability of free space (~ $4\pi \cdot 10^{-7}$ H/m)
ε	Permittivity (when subscripted, *d* stands for dielectric)
ε_0	Permittivity of free space (~ $8.854 \cdot 10^{-12}$ F/m)
β	Phase constant
ρ_0	Port size (radial waveguide method)
W	Port lateral size (cavity resonator model)
ϕ	Potential
γ	Propagation constant
v_p	Propagation velocity
Q	Quality factor
μ_r	Relative permeability
ε_r	Relative permittivity
R	Resistance
t_r	Rise time
s	Separation
t_s	Skin depth
S	S-parameter
t	Time
Δl	Trace length mismatch
T_i	Transformation matrix for currents (modal decomposition)
T_v	Transformation matrix for voltages (modal decomposition)
l	Transmission line/ trace length
Y^v	Via admittance (matrix)
r^{ap}	Via-antipad radius
Z^{v0}	Via impedance
r^p	Via-pad radius
r^v	Via radius
Y^c	Via-to-plane admittance
C^v	Via-to-plane capacitance
C^{vb}	Via-to-plane capacitance of a buried via
C^c	Via-to-plane coaxial capacitance

C^f	Via-to-plane fringing capacitance
Z^c	Via-to-plane impedance
C^b	Via-to-plane lateral capacitance
V	Voltage
η	Wave impedance
$\underline{k}$	Wave number
λ	Wavelength
Y	Y-parameter
Z	Z-parameter

Acronyms

2D	Two Dimensional
3D	Three Dimensional
ABC	Absorbing Boundary Condition
ABCD	Microwave Network Chain Parameters
AC	Alternating Current
BGA	Ball Grid Array
C	Capacitance
CIM	Contour Integral Method
CISPR	International Special Committee on Radio Interference (in French)
CRM	Cavity Resonator Model
CM	Common-Mode
CRM-DS	Cavity Model – Double Summation
CRM-SS	Cavity Model – Single Summation
CRM-SSi	Cavity Model – Single Summation (improved)
CPU	Central Processing Unit
DC	Direct Current
DDR	Dual Data Rate (for Random Access Memories)
Decap	Decoupling Capacitor
DM	Differential Mode

DR	Data Rate
DVI	Digital Visual Interface
EM	Electromagnetic
EMC	Electromagnetic Compatibility
EMI	Electromagnetic Interference
ESL	Equivalent Series Inductance
ESR	Equivalent Series Resistance
FCC	Federal Communications Commission
FDTD	Finite Difference Time Domain Method
FEM	Finite Element Method
FFT	Fast Fourier Transform
FIT	Finite Integration Method
FSV	Feature Selective Validation
FWHM	Full Width Half Maximum
GND	Ground
GS/SG	Ground Signal / Signal Ground
$h-$	Microwave Network Hybrid Parameters
HTCC	High Temperature Co-fired Ceramic
HFSS	High Frequency Structure Simulator (FEM Solver)
IBM	International Business Machines Corp.
IC	Integrated Circuit (or chip)
IFFT	Inverse Fast Fourier Transform
ITRS	International Technology Roadmap for Semiconductors
LAN	Local Area Network
LGA	Land Grid Array
LTCC	Low Temperature Co-fired Ceramic
MAN	Metropolitan Area Network
MCM	Multi-chip Module
M-FDM	Multilayered Finite Difference Method
MoM	Method of Moments
MTL	Multiconductor Transmission Line
PC	Personal Computer

PCB	Printed Circuit Board
PCIe	Peripheral Component Interconnect Express
PDN	Power Distribution Network
PEC	Perfect Electric Conductor
PEEC	Partial Element Equivalent Circuit Method
PI	Power Integrity
PLL	Phase-Locked Loop
PMC	Perfect Magnetic Conductor
PML	Perfectly Matched Layer
PWR	Power
RAM	Random Access Memory
RE	Radiated Emissions
RLC	Resistance-Inductance-Capacitance
RPL	Recessed Probe Launch
RW	Radial Waveguide
RW-IT	Radial Waveguide – Image Theory
Rx	Receiver
S-	Microwave Network Scattering Parameters
S[X]	Signal Level [X]
SA-SCSI	Serial Attachment – Small Computer System Interface
SATA	Serial Attachment (Interface)
SBU	Sequential Build-Up
SerDes	Serializer-Deserializer
SI	Signal Integrity
SiP	System in Package
SMT	Surface Mount Technology
SoC	System on Chip
SOLT	Short Open Load Thru (Calibration Procedure)
SoP	System on Package
SPICE	Simulation Program with Integrated Circuit Emphasis
SSN	Simultaneous Switching Noise
TE	Transfer Electric

TEM	Transfer Electric Magnetic
TET	Institut für Theoretische Elektrotechnik, TUHH
TL	Transmission Line
TLM	Transmission Line Matrix Method
TM	Transfer Magnetic
TMM	Transmission Matrix Method
TSV	Thru Silicon Vias
TUHH	Technische Universität Hamburg-Harburg
TV	Test Vehicle
Tx	Transmitter
USB	Universal Serial Bus
VNA	Vector Network Analyzer
VIA	Vertical Interconnect Access
VPF	Via Pin Field (Simulation Tool)
VRM	Voltage Regulator Module
WAN	Wide Area Network
Y-	Microwave Network Admittance Parameters
Z-	Microwave Network Impedance Parameters

1. Introduction

1.1. Motivation and Context of this Work

Modern high-speed electronic systems often require thousands of off-chip interconnects to interface heterogeneous components such as processing units, memory, storage devices, and network interfaces. The trends towards miniaturization and higher data rates that have driven the electronic industry in the past few decades demand high-density interconnects operating in the multi-GHz range. For instance, multi-chip-modules (MCM) and high-performance printed circuit boards (PCBs) are commonly found in commercial products [1]-[2]. Moreover, several high-speed specifications for wired links –such as PCIe, SATA, DDR-3, or 10G Ethernet– have become industry standards, many of them targeting data rates in excess of 10 Gigabits per second (Gb/s) [3]-[4] (Figure 1.1).

Off-chip interconnects constitute the bottleneck for the maximal achievable data rate, since they introduce frequency dependent degradation and distortion on signal paths [5]. The efficient modeling and simulation of off-chip interconnects have become essential to assist the design process and to look for the best trade-off between cost and performance. This is a challenging task because of the large number of elements and the complicated electromagnetic field effects that must be considered to model a realistic scenario. General-purpose numerical methods can be used to describe interconnects accurately, but often become inefficient to handle complex configurations and to perform trade-off and optimization studies. In contrast, simplified quasi-static models usually fail to describe the high-frequency behavior and the multiple coupling mechanisms of interconnect systems. According to the international technology roadmap of semiconductors ITRS 2009 [6], the development of accurate and efficient compact models for high-frequency circuits and systems is a challenge of prime importance in order to enable the concurrent design and optimization of integrated circuits (ICs), passives, and substrates.

In the category of off-chip interconnects, multilayer substrates, often used in packages and PCBs, constitute an important resource for integration of digital, analog, and passive components. Nowadays, this type of substrate can be found in almost any

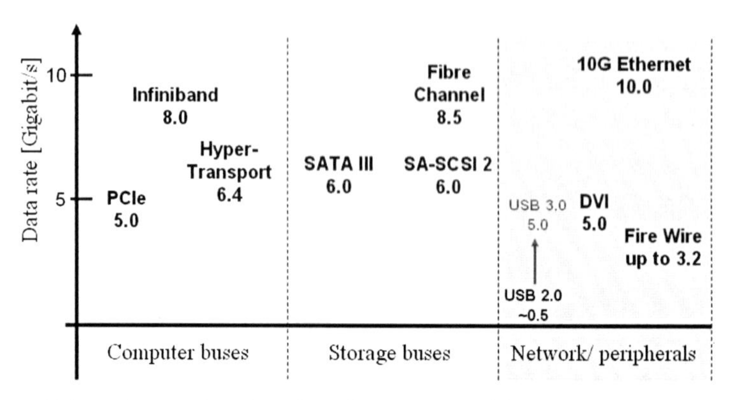

Figure 1.1 Some high-speed standards for digital systems and their maximal data rates (in 2009).

electronic system, from hand-held devices to mainframe servers, communication, and space applications [7].

This work is a contribution to the topic of efficient modeling and simulation of multilayer substrates. Semi-analytical models for the electrical behavior of vias and traces in parallel-plane environments have been proposed and validated against other numerical techniques and measurements. The models have been incorporated into an automated simulation method that has been successfully applied to signal integrity, power integrity, and electromagnetic interference analyses in a comprehensive and very efficient manner. At the present time, it is possible to handle relatively complex structures with hundreds of signal and power/ground vias, tens of mixed reference planes, several coupled traces, and lumped elements such as decoupling capacitors. The utilization of these models allows a reduction of the computation time by a factor between two and three orders of magnitude when compared to general-purpose numerical methods for electromagnetic field simulation.

1.2. Organization of the Work

The bulk of this thesis is organized into six chapters, whose contents are detailed as follows.

Chapter 2 presents a very brief introduction to multilayer substrate technologies. The main concerns in relation to signal integrity, power integrity, and electromagnetic compatibility are reviewed and references for further reading are provided. The state of the art methods for modeling and simulation of multilayer substrates are overviewed in the last subsection.

Chapter 3 reviews the physical effects associated with vias, which are fundamental for understanding the field coupling mechanisms in multilayer substrates. The addressed topics cover the excitation of parallel-plate modes, effects of return vias, and crosstalk. The extraction of via transmission line parameters, the impact of via geometry, and the stub effect are discussed as well.

Chapter 4 explains the modeling approach, the proposed models, and their mathematical formulation in terms of microwave network parameters. The via model and different alternatives to compute its building blocks are addressed in this chapter. Analytical techniques to compute the parallel-plate impedance are reviewed, compared, and a hybrid method is proposed. Alternatives to compute the via-to-plane capacitance and the extension of the via model to consider traces by applying a modal decomposition technique are also covered. Finally, the approach used for partial result concatenation and the general simulation framework for multilayer substrates are introduced.

Chapter 5 deals with the validation of the models. The presented examples cover the simulation of multilayer via configurations, single-ended and differential links, and via arrays. The described structures include signal, power, and ground vias with mixed-reference planes, buried and blind via configurations. The results are compared against general-purpose full-wave solvers and hardware measurements. The numerical efficiency of the models is also discussed and, in the last section, model limitations and perspectives for further developments are provided.

Chapter 6 presents the application of the models to signal integrity, power integrity, and electromagnetic interference problems. The case studies cover the simulation of differential links across ball grid via arrays (BGA), the via stub effect, and mode conversion in differential links. The co-analysis of power and signal integrity, and the impact of power/ground via design and surface decoupling capacitors are discussed next. In the last section, the extension of the method to the combined simulation of signal propagation, ground bounce, and radiated emissions is explained.

Chapter 7 gathers the main results of this work. The contributions of this thesis are briefly reviewed and pros and cons of the modeling approach are discussed. Recommendations for further work are also provided.

1.3. Conference and Journal Contributions

As part of this work several conference [8]-[16] and journal contributions [17]-[19] have been made in the field of model development, efficient simulation, and analysis of multilayer substrates. These publications constitute a fundamental part of this thesis. During the development of this project the author has also contributed to a number of publications in related topics [3],[20]-[27].

Much of this work has been done in cooperation with industry and other university research groups. The main external collaborators were the High-Speed I/O Subsystems and Packaging Group at IBM T. J. Watson Research Center, who also provided the test hardware and most of the measurements presented in Chapters 5 and 6, and the Electromagnetic Compatibility Laboratory at the Missouri University of Science and Technology.

2. Multilayer Substrates in High-Speed Electronic Systems

The main task of interconnects is to allow the flow of information between diverse components of an electronic system. They also provide the mechanical support and the interface for heterogeneous technologies [7],[29]. Interconnects are necessary at all levels of hierarchy, from integrated circuits up to overseas communication links. Optical technologies are ubiquitous in very long distance applications; however, short interconnects are still the domain of wired links using metallic conductors, typically Copper (see Figure 2.1). Although the realization of short optical links has been demonstrated, e.g. in [30], important cost and fabrication obstacles have to be overcome before their widespread adoption [31]. For this reason, the improvement of wired technologies through the miniaturization of systems and the development of global and more efficient modeling strategies are currently active research fields [32]-[33].

As depicted in Figure 2.2 for systems using "short" electrical links, interconnects can be classified into two main categories:

- *On-chip*, formed by the metallization layers of IC technologies and pads that provide the interface to the next packaging level. For dense digital circuits they are typically organized as grid arrays [34].
- *Off-chip*, defined by the first, second, and third packaging levels. They comprise multilayer substrates at package and board levels, as well as other elements such as bond wires, solder balls, pins, connectors, sockets, cables, etc. [29]. This second category is the one of interest for this work and it will be briefly discussed in the next section.

2.1. Multilayer Substrate Technologies

Multilayer substrates are commonly formed by stacked power/ground levels and signal layers that are linked to each other by vertical interconnects, called *vias* (Figure 2.2).

Laminated substrates are fabricated by alternating *core* and *prepreg* layers. The core is a cured dielectric layer bounded by two metallic plates, whereas the prepreg is an uncured dielectric material often consisting of woven glass sheets impregnated with

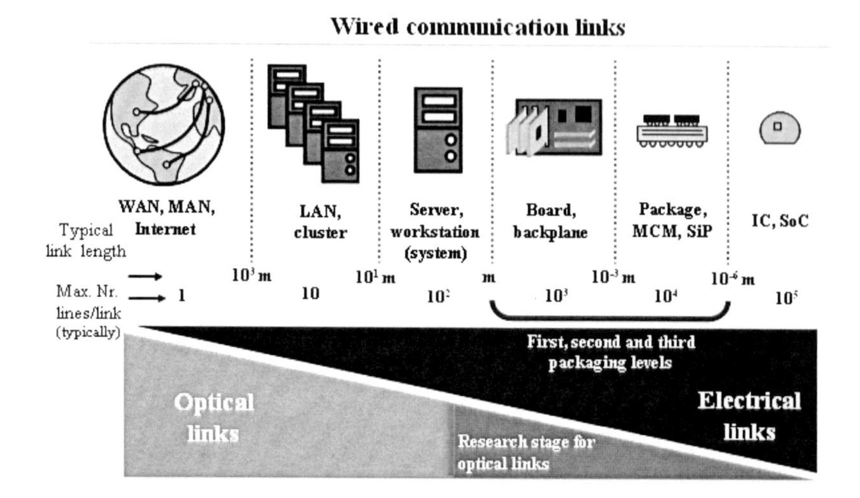

Figure 2.1 Utilization of optical and electrical links for high-speed communications and digital systems. Adapted from [31].

epoxy resin. The prepreg is used to bond multiple stacked cores in a single step by applying heat and pressure. The final arrangement of cores and prepreg layers is known as *stackup* [29]. The circuitry is formed by etching conducting patterns on the metal plates prior to lamination and interconnecting the different levels by means of vias [35]. Vias are usually created by drilling holes spanning the full stackup and subsequently plating a metal (typically Copper) on the walls.

Sequential built-up processes (SBU, also called build-up) are used for high performance boards that require higher densities or drastic geometry size changes. In SBU each layer is processed separately. The vias, called *microvias*, can be formed by photo-processes, etching, or laser drilling [35]. SBU allows the fabrication of smaller interconnect elements in comparison to laminated processes, such as buried microvias.

A multilayer substrate may contain both laminated and SBU regions, as depicted in Figure 2.3. Bond wires or solder balls are used to interface the package with ICs, whereas solder balls or pins may serve to reach the next package level. Metallic *pad* regions are used to interconnect vias with other elements such as traces, decoupling capacitors, or solder balls. Clearance holes, known as *antipads*, are etched into the planes in order to isolate vias when connectivity to the plane is not desired.

First level packages can be made with organic materials (e.g. epoxy resins, polymides), ceramics (e.g. HTCC, LTCC), or other plastic and flex-film compounds [7].

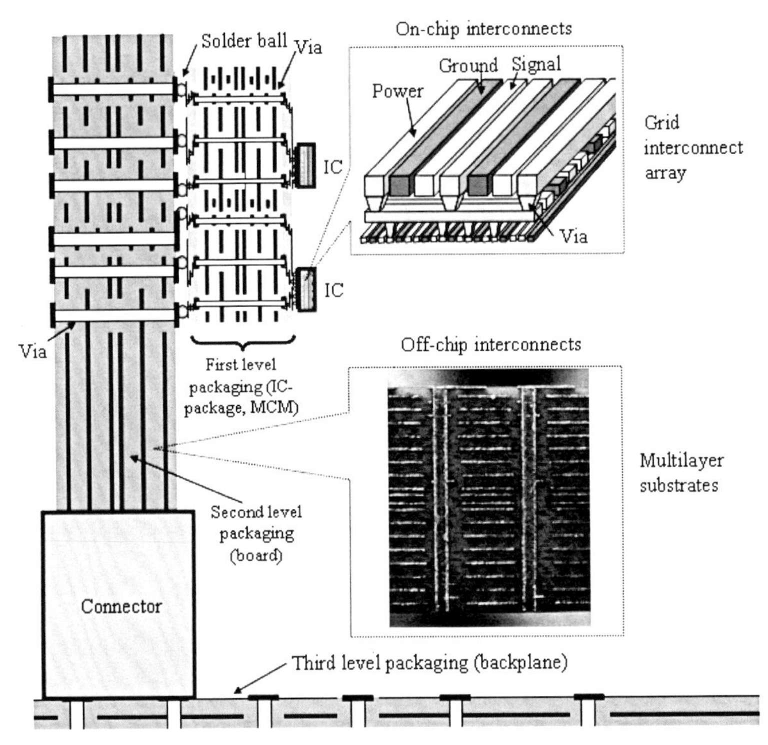

Figure 2.2 Illustration of interconnect levels in high-speed electronic systems. On-chip interconnects mainly consist of metallization layers of IC technologies, which are typically arranged as grid meshes. Off-chip interconnects cover the first (package and MCM), second (board), and third (motherboard/ backplane) levels. Multilayer substrates are used to support the off-chip signal and power networks.

Organic packages offer a low-cost solution, whereas ceramic packages provide better mechanical performance and higher wiring and pin densities [31]. With SBU processes, microvias can have diameters below 50 microns (μm) and center-to-center separations (pitch) starting from 100 microns. MCM modules and chip stacking are other available package technologies that allow the integration of several ICs, including processors and cache memory. Silicon carriers with through silicon vias (TSV) have been reported in the literature as an alternative which allows stacking of multiple dies (3D integration or System-in-Package) [32],[36]. TSV diameters are typically in the micron (μm) range, from 0.1 up to several tens of microns [36].

Similarly to low-cost packages, conventional PCBs are fabricated by lamination processes. High density boards may have over 25 metallic layers and can be made with a wide variety of dielectric materials and copper-based foils. Reinforcement cloths can also be incorporated in the dielectric regions in order to provide dimensional stability

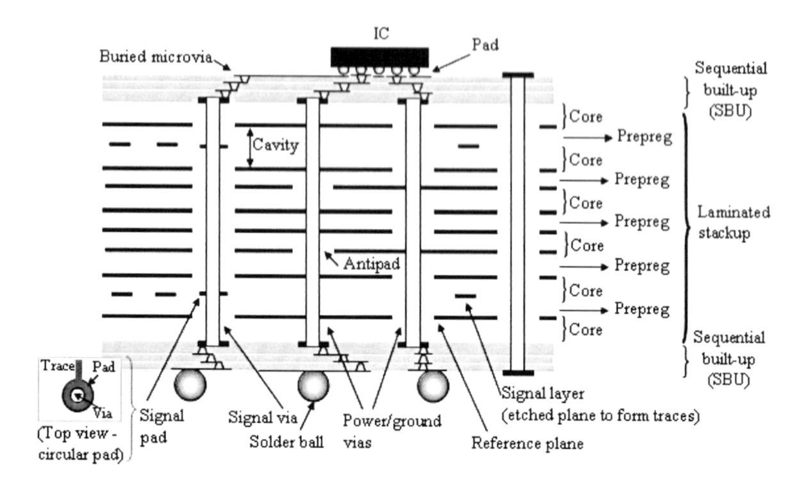

Figure 2.3 Diagram of a multilayer substrate combining a laminated core with sequential built-up layers.

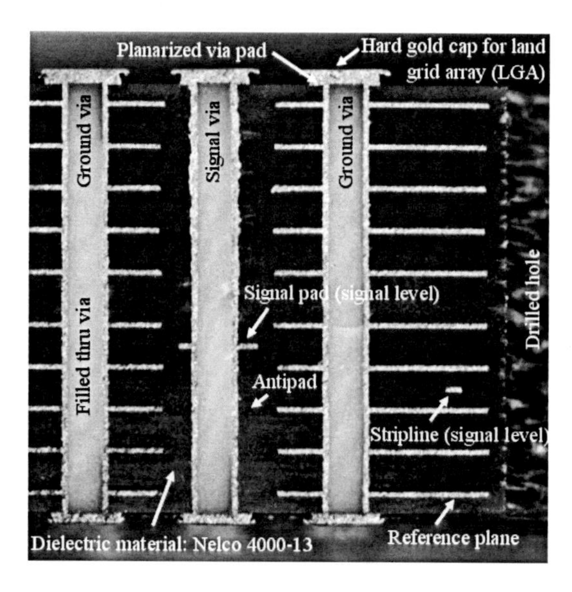

Figure 2.4 Multilayer printed circuit board (PCB) cross section. All reference planes are connected to ground vias and two signal levels can be identified on the cut. *Photo courtesy of Y. H. Kwark, IBM T. J. Watson Research Center, NY, USA.*

during reflow and rework processes [37]. Typical separations between metallic levels can be as low as 2.5 mil (1 mil ≈ 25.4 μm) and up to 30 mil. Through-via diameters for high-performance applications range between 8 and 16 mil (~ 0.2 - 0.5 mm), with antipad diameters from 15 to 40 mil and a via pitch of around 40 mil (~ 1mm). With SBU higher densities can be achieved. Microvias can have diameters below 2 mil [38]. Figure 2.4 shows a photograph of a multilayer PCB cross-section. Available processes and technologies for packages and PCBs are widely discussed in several books, for instance in [7],[37]-[38].

The present work is focused on the efficient modeling of the portion of the substrate enclosed between solid reference planes, which are affected by the excitation of parallel-plate modes [19]. The fact that at high frequencies the vias and the reference planes become tightly coupled has significant signal integrity (SI), power integrity (PI) and electromagnetic compatibility (EMC) implications. The next sections briefly review the main concerns from each one of these three perspectives in relation to the reliable design of off-chip interconnect systems.

2.2. Signal Integrity

Signal integrity is concerned with the reliable transmission of information across the different interconnect hierarchy levels [39]-[40]. Passive wired interconnects are band limited, typically showing a low-pass characteristic. It is therefore important to ensure that the channel is able to transmit necessary frequency components of the signals without excessive degradation. The bandwidth of digital signals depends on the rise time t_r and it can be estimated from the analysis of idealized waveforms [39] (See Appendix A.1). For a given data rate (DR), the bit time is $T_b = 1/DR$ and the fundamental frequency is $f_s = 1/T = 1/(2T_b)$. A typical t_r value is about $0.1 \cdot T$ and the maximum frequency content of the signal can be approximated as $f_{max} \approx 1/t_r$. For instance, for a 10 Gb/s signal, the fundamental frequency is 5 GHz, $t_r \approx 20$ ps, and $f_{max} \approx 50$ GHz. A less stringent criterion which uses $0.5/t_r$ to include about 90 % of the frequency content yields $f_{max} \approx 25$ GHz [41]. These quick estimates allow judging the interconnect performance and the required bandwidth of the models used to describe the channel.

A signal transmitted over off-chip interconnects can suffer degradation (in magnitude and phase) due to [5],[42]:

- Frequency dependent dielectric and conductor loss.
- Reflections at discontinuities such as vias and plane perforations.
- Noise sources such as crosstalk from other signal nets as well as switching noise coupled in through the power distribution network.

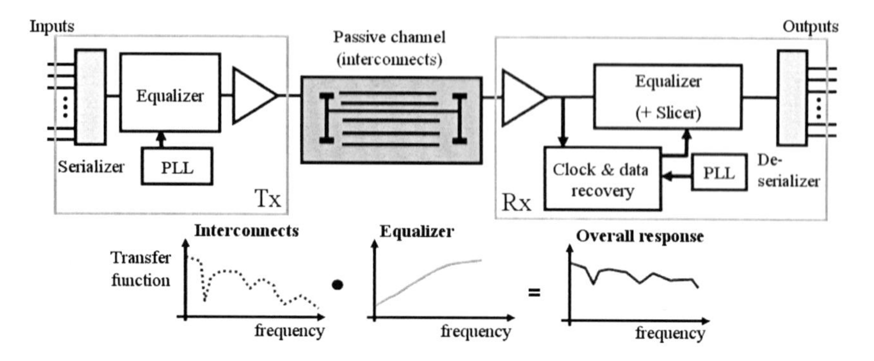

Figure 2.5 Simplified diagram of a high-speed serial link (SerDes). Equalizers at transmitter (Tx) and receiver (Rx) sides are used to compensate the channel degradation at higher frequencies. A more detailed treatment of the topic can be found in [42]-[44]. Heuristically, the task of the equalizers is to compensate the low-frequency response of the channel by emphasizing the high-frequency components of the signals. PLL stands for phase-locked loop and it is used to provide an stable clock reference.

These effects can be mitigated by proper design of the passive channel, which includes the selection of the substrate, stackup, via types and position, number and position of ground and power vias, routing of signal paths, type of signaling, utilization of decoupling capacitors, etc..

For links working at multi-gigabit data rates, active digital circuits are frequently required to further compensate the channel degradation (Figure 2.5). Equalization techniques are used to "flatten" the frequency response of the channel by counterbalancing its low-pass characteristic by means of "high-pass" analog and digital filter operations [42]-[44]. Nevertheless, additional circuits are expensive in terms of area and power consumption, and therefore the optimization of the interconnects aiming at a reduction of the necessary on-chip circuitry is highly desirable.

2.3. Power Integrity

The function of the power distribution network (PDN) is to provide a clean and reliable supply voltage to all system components and a good return path for signals. The most important components of a PDN are the main DC supply source, voltage regulator modules (VRM or DC-DC converters), decoupling capacitors, and the associated interconnects [45]. At package and board levels, the PDN interconnects are mostly formed by the reference planes, and the ground and power vias used to link them (Figure 2.6). Although power meshes can also be used at package level [46], stacked solid planes are usually preferred since they can provide a lower PDN impedance and the inter-plate capacitance may help to reduce crosstalk and electromagnetic interference.

The voltage and current variations on the PDN induce power noise through the parasitics associated with on- and off-chip interconnects, which may affect the amplitude (under/over shot) and delay (jitter) of information signals. It is therefore desirable to control the PDN impedance to get very small impedance values. In fact, modern integrated circuit technologies demand target impedances in the milli-Ω range [45].

The simultaneous switching noise (SSN) [47], also called Δi noise, is perhaps the most important mechanism for power noise generation. This phenomenon takes place when many digital circuits change states simultaneously and the associated peak in the current demand induces a transient voltage drop on the PDN. This voltage is, in a first order approximation, proportional to the equivalent inductance of the interconnects L_{eq}

$$v_{ind}(t) = L_{eq} \cdot \frac{di(t)}{dt} \tag{2.1}$$

Unlike other noise sources, the SSN is deterministic since it can be related to the period of the signal transitions [39]. This type of noise can be either generated on chip, due to IC core circuits and drivers switching, or off chip, due to signals traveling on signal nets at package and board levels. The power noise can be mitigated by [39]:

- Limiting the rise time of the signals in order to reduce the di/dt magnitude (which is in conflict with signal integrity interests of high data rates and signal detection).

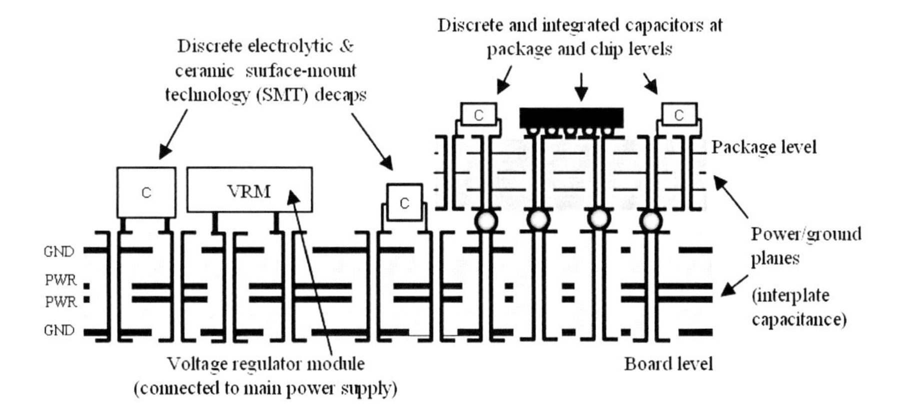

Figure 2.6 Diagram of a power distribution network main components at board and package levels. Voltage regulator modules and decoupling capacitors (C) are used to provide a reliable power supply.

- Reducing the interconnect inductance. This is achieved for off-chip PDNs by using solid power and reference planes located in close proximity and by optimizing the placing of PDN components.
- Bypassing the PDN impedance with decoupling capacitors.

While the VRM provides a large charge reservoir for the circuits, it can not always meet the demands quickly enough due to the parasitic inductance of the PDN. In this time domain picture, decoupling capacitors serve as additional charge sources that can be placed closer to the chips and help to maintain the voltage level at the moment when several gates switch and the current demand drastically increases [48]. In the frequency domain, the task of these capacitors can be seen as the reduction of the equivalent PDN impedance by providing a low impedance path between power and ground nets (bypassing). However, non-ideal decoupling capacitors possess parasitic inductance and resistance, which limit their practical effectiveness [49]. Larger capacitors can provide more charge but have larger parasitics –both internally and due to their associated interconnects– that make them effective only at lower frequencies. Smaller capacitors can be placed closer to the problematic nets and are more effective at higher frequencies. The decoupling scheme in modern electronic systems requires many capacitors placed at all levels of the interconnect hierarchy, from electrolytic and surface capacitors at board and package levels up to embedded package and on-chip capacitors [45].

2.4. Electromagnetic Compatibility

Electromagnetic compatibility addresses the design of electronic systems that do not cause electromagnetic interference (EMI) to other devices, and are not susceptible to EMI coming from other systems [50]. The EMI emissions can be *conducted* and pass to the common AC power net, or they can be *radiated* directly from the device. The conducted emissions can be transformed to radiated emissions at the power nets [51].

Devices and interconnects placed on multilayer substrates, such as microstrip lines or plane discontinuities, can become a source of EMI. Radiated emissions can also be originated at the board edges, in particular at higher frequencies. Other sources of interference are ICs and cables attached to boards. The mitigation of emissions can be achieved either by suppression of the EMI source, elimination of the coupling paths, or reduction of the susceptibility of the devices by increasing noise margins [50].

Emissions are regulated by standards set by governmental agencies, such as the Federal Communications Commission (FCC) in the United States and the International Special Committee on Radio Interference (CISPR) in the European Union [50]-[51]. These standards attempt to control EMI pollution, but do not guarantee the functionality of a system; therefore, manufacturers may impose additional requirements, particularly for devices working in the GHz range.

SI, PI, and EMC are three closely related, yet different aspects of the electromagnetic nature of digital interconnects. For instance, common-mode currents of signal nets and SSN are important sources of radiated emissions [50]. A good design methodology should be able to address the interdependences between these different perspectives.

2.5. Overview of Techniques for High-Frequency Modeling of Multilayer Substrates

Analytical approximations, simplified static and quasi-static models for interconnect elements –like some formulae provided, for instance, in [41],[52]-[53]– have proven to be useful in developing understanding and performing low-frequency analyses. However, the simulation of SI and PI effects associated with high-speed systems requires more advanced techniques that are able to account for non-quasi-static field effects and deal with complex coupling mechanisms among many elements. General purpose numerical techniques, e.g. full-wave methods, have been used in the past for this purpose. Some examples are the finite difference time domain method (FDTD) [54], the partial equivalent electric circuit method (PEEC) [55], and the finite element method (FEM) [56]. Commercial tools that are based on the aforementioned techniques are widespread. Although full wave solvers provide the best flexibility to handle arbitrary geometries, the main disadvantage of these methods is that a full discretization of the model is required and therefore the computational burden rapidly grows as the operating frequency, size, and complexity of the interconnect structure increase.

For this reason, much effort has been expended to develop customized numerical or semi-analytical methods for analysis of multilayer substrates. Most of these approaches exploit the planar nature of the PDN [57]-[58], since the cavities, formed between adjacent reference planes, are in general very thin in comparison to the wavelengths of interest. Some of the techniques have been applied to the analysis of via transitions and power planes such as the multilayered finite-difference method (M-FDM) [59], the contour integral method (CIM) [60]-[62], the transmission matrix (TMM) [63] and transmission line matrix method (TLM) [64], multiple scattering methods [65]-[68], as well as analytical formulations [69]-[71].

The available methods can be further improved by offering higher flexibility and better system-level simulation capabilities, which is useful to enable the design, analysis, and pre-layout optimization of very complex structures at different levels of hierarchy (IC, package, board, system) and from different perspectives (SI, PI, EMC). It is also desirable to achieve these goals without the expenditure of massive

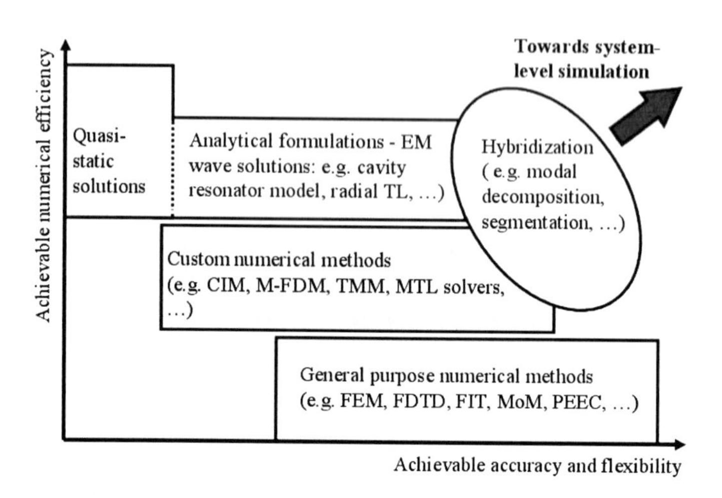

Figure 2.7 General classification of available methods for high-frequency modeling of multilayer substrates.

computation. This demands more efficient and compact high-frequency interconnect models, and a better understanding of their applicability, limitations, and related techniques to compute their constituent elements.

In this vein, several hybrid techniques for analysis of complex multilayer structures, combining some of the aforementioned approaches, have also been proposed, for example in [72]-[77]. Segmentation techniques [60],[78] have been applied to describe multilayer power planes, and the modeling of traces connecting vias has been included using modal decomposition methods [79]. In fact, much of the work related to this thesis is a hybridization of diverse analytical and numerical techniques [19], where the co-simulation of SI, PI, and electromagnetic interference is also addressed [12],[17]. Figure 2.7 illustrates the different approaches mentioned above, classified in terms of computational efficiency and flexibility to handle arbitrary configurations and geometries.

3. Physical Effects Associated with Vias

As introduced in Chapter 2, multilayer substrates require the utilization of vias to interconnect signal traces and reference planes laid out at different levels. In contrast to the transmission line theory used to describe traces, the problem of via modeling is more complex due to its intricate environment [80]. These vertical interconnects act as discontinuities that couple signal and power nets, which lead to signal, power integrity, and other electromagnetic interference problems. Proper via design is therefore essential for interconnect systems capable of operating at Gb/s data rates [81].

At low frequencies, typically in the MHz range, quasi-static models are useful describing the behavior of vias [82]. However, at higher frequencies, the via currents excite parasitic modes between the reference planes, which are associated with wave propagation effects that cannot be captured with a quasi-static approach [54],[80]. Consequently, via modeling in the GHz range requires a good understanding of the underlying electromagnetic mechanisms.

In this chapter, the main physical effects associated with vias are reviewed, including the excitation of parallel-plate modes, the role of ground vias, and crosstalk. The extraction of transmission line parameters for vias, the influence of their geometry, and the stub effect are also discussed with the help of a set of generic examples.

3.1. Types of Signal Vias

Common types of vias for utilization in multilayer substrates are illustrated in Figure 2.1. They are classified according to the topology of signal and reference levels that are part of the transition. A through-hole via (or simply thru) goes from side to side of the entire stackup. The via segment under or over stripline transitions is called via stub and it may insert unwanted resonances. This phenomenon is called the stub effect and it will be discussed in Section 3.6. Although the most commonly utilized due to its low fabrication cost, the thru configuration is detrimental in terms of crosstalk and because of the stub effect. For this reason, other types of vias can be fabricated at the cost of a more complex and expensive manufacturing process. A blind via passes through a section of the stackup and it is only visible from one board side. Thru vias can be

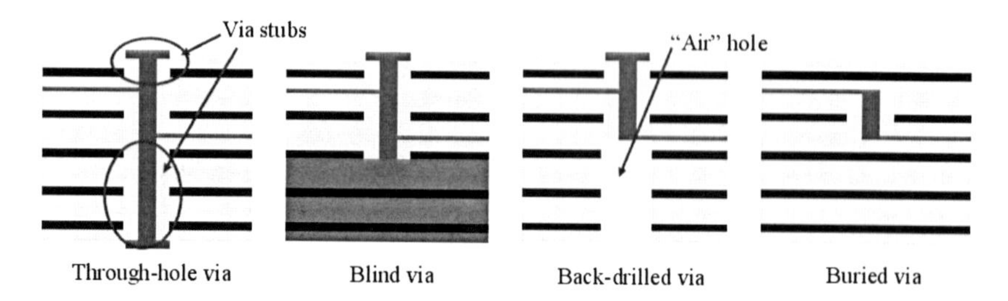

Figure 3.1 Common via types used in multilayer substrates.

converted to blind ones by removing their stub sections in a post-processing step known as back-drilling. Blind and buried vias, which only cross some of the internal cavities of the board, can also be fabricated with sequential processes (i.e. SBU, see Section 2.1).

The examples studied in this chapter illustrate thru via behavior; the basic structure is depicted in Figure 3.2. These were simulated with a full-wave solver utilizing the finite integration technique (FIT) in time domain [83].

3.2. Excitation of Parallel-Plate Modes

A typical via crosses many layers and the currents flowing on it, perpendicular to the signal and reference levels, can excite parasitic modes in the cavities formed between adjacent reference planes. They are called *parallel-plate modes.* Unlike conventional transmission lines, the return path for via currents is more complex and is influenced by the electromagnetic coupling between the plates and the number and position of return vias (Figure 3.2(a)).

Assuming that the via currents are uniformly distributed, the parallel-plate modes are, in principle, guided electromagnetic waves of cylindrical symmetry [84]. Because of the small separation between reference planes, it can also be assumed that only TEM waves are supported inside the cavities and that these fields are constant in the perpendicular direction (z-axis). Each cavity is then treated as a radial transmission line, with electromagnetic fields defined in cylindrical coordinates as [85]

$$\mathcal{E}_z(\rho,t) = A \cdot H_o^{(1)}(\underline{k}\rho) + B \cdot H_o^{(2)}(\underline{k}\rho)\,, \tag{3.1}$$

$$\mathcal{H}_\varphi(\rho,t) = \frac{j}{\eta} \cdot \left(A \cdot H_1^{(1)}(\underline{k}\rho) + B \cdot H_1^{(2)}(\underline{k}\rho)\right), \tag{3.2}$$

with $H_n^{(1)}$, $H_n^{(2)}$ the Hankel functions of order n, of first and second kind, respectively. The terms A and B comprise the complex amplitude and the harmonic time dependence of the inward and outward traveling waves, respectively, $\underline{k} = \omega\sqrt{\mu\varepsilon}$ is the

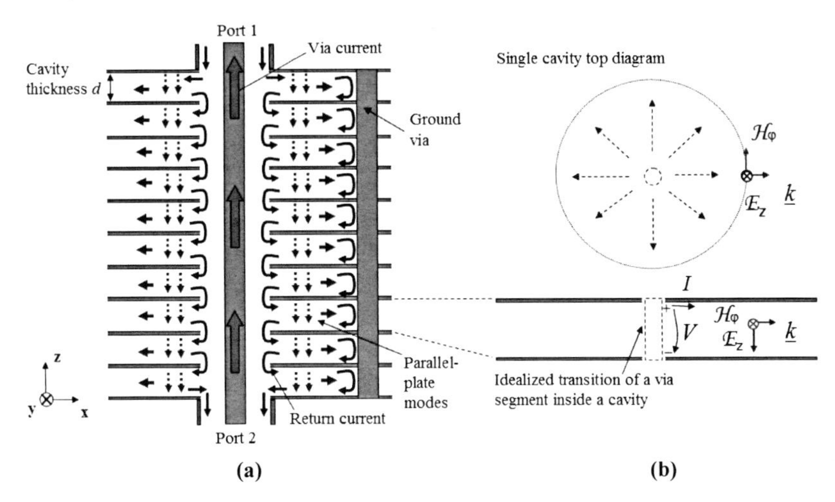

Figure 3.2 Illustration of parallel-plate modes excited by via transitions. (a) Signal and return path for the currents of a multilayer via. The return path is shared by the return vias and the reference planes, which can support propagating waves. (b) Definitions for a single via segment in a cavity, assuming infinite planes and neglecting the near field in the antipad region.

wave number, and $\eta = \omega\mu / \underline{k} = \sqrt{\mu / \varepsilon}$ the wave impedance. The first term in Eqs. (3.1) -(3.2) represents radial waves traveling inward, whereas the second depicts outward traveling waves. If it is assumed that the reference planes extend to infinity, the inward traveling wave terms vanish. The terminal voltage and current for the radial transmission line are written as (Figure 3.2(b))

$$V = -\mathcal{E}_z \cdot d \,, \tag{3.3}$$

$$I = 2\pi r \cdot \mathcal{H}_\varphi \,. \tag{3.4}$$

The origin of the local coordinate system is defined at the via centers, with the radial coordinate $\rho = (x^2+y^2)^{1/2}$. Higher order TM and TE modes in z-direction inside the cavities could be also supported, however they have cutoff frequencies given by [54],[85]

$$f_c = \frac{n}{2d\sqrt{\mu\varepsilon}} \quad \text{with} \quad n = 1,2,3,... \tag{3.5}$$

with d the separation between power/ground plates. For a maximum frequency of 40 GHz and $\varepsilon_r = 4$, d must be larger than 70 mil to support the propagation of higher order modes. In practice, this separation –typically between 10 and 30 mil for high-speed PCBs– is often much smaller than the minimum wavelength of interest and it locates the cutoff frequencies of higher order modes far away (beyond 100 GHz). These evanescent modes, however, can influence the local via environment as it will be discussed next.

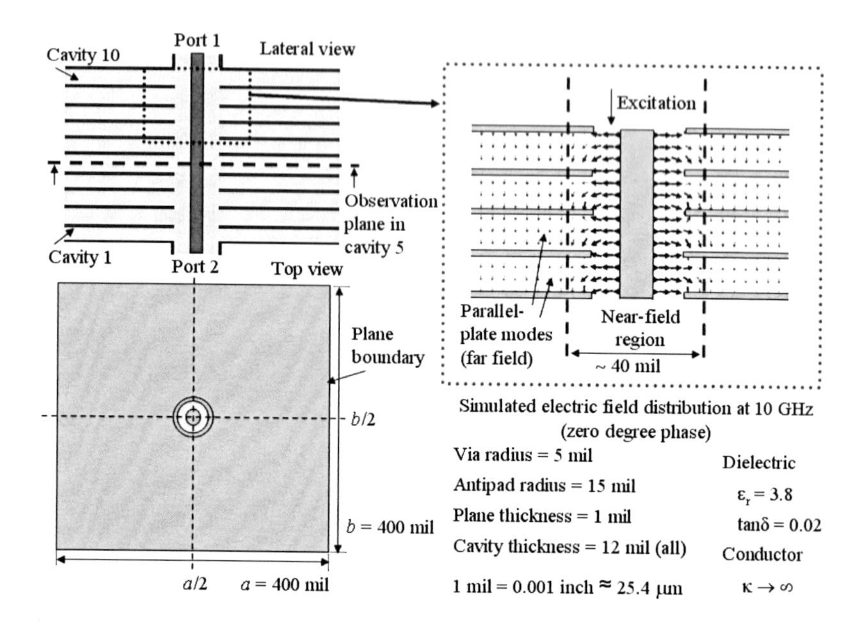

Figure 3.3 Example of a multilayer via. The structure was simulated with the finite integration technique (FIT) [83] assuming perfectly matched layer (PML) boundaries at the board edges. The simulated electric field vector distribution on the cross section of the via shows the presence of parallel-plate modes and inhomogeneous fields in the antipad region. The excitation is a normalized Gaussian pulse (f_{max} = 40 GHz) applied to port 1.

The model for the full-wave simulation of a multilayer via is described in Figure 3.3. Initially, it is supposed that the planes are infinitely large, which means that there are no reflections coming from board edges and only outward traveling waves need to be considered. This condition is numerically modeled by an absorbing boundary condition (ABC) or perfectly matched layers (PML). Two wave ports are defined at both via ends and the excitation is a 1-$(W)^{0.5}$ Gaussian pulse with a maximum frequency content of 40 GHz. The time domain results are converted, in a post-processing step, into the frequency domain to get the field distributions and *S*-parameters. The coaxial extensions on via sides are necessary to provide a regular cross-section for the wave ports in the full-wave simulation.

The numerical field simulation in Figure 3.3 shows the excitation region where the parallel-plate modes start to develop. Figure 3.4 illustrates the cylindrical nature of the parallel-plate waves by plotting the magnitude of the electric field inside the fifth cavity at different frequencies. The excitation was applied at the port number one, located at the top side of the via.

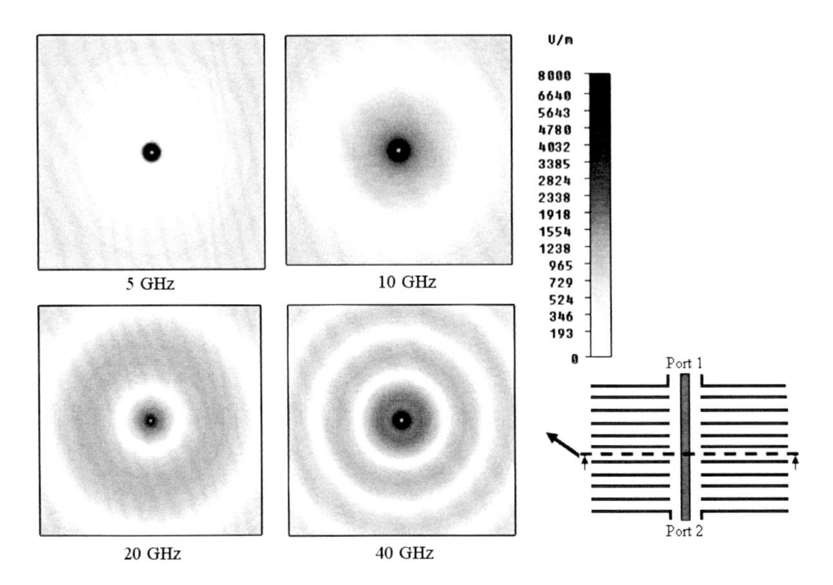

Figure 3.4 Magnitude of the electric field inside the fifth cavity of the structure in Figure 3.3, for zero degree phase and assuming PML boundary condition at the board edges. The plots show the simulated cylindrical waves guided between two reference planes at different frequencies. The excitation is applied to port 1 and the fields are plotted on a logarithmic scale.

The vector field pattern in Figure 3.3 also indicates that an inhomogeneous field region exists near the via barrel and the antipad region, which has a diameter of about 40 mil in the case shown. These near fields can be related to the non-propagating modes inside the cavities. Their effect can be approximated with lumped capacitances for many practical situations (see Section 4.4). Near-field coupling is possible for vias in very close proximity or for vias sharing the same antipad such as differential ones.

For finite planes, outgoing waves are reflected at the board edges. An open boundary condition, also denoted as perfect magnetic conductor (PMC), usually serves as a good approximation given the small separation between the planes. However, at higher frequencies, particularly for low loss and thick cavities, the final impedance of the space surrounding the board edges may become important and lead to a considerable amount of radiated emissions [17]. The plots in Figure 3.5 show the resultant electric field distribution in presence of ideal reflective open boundaries. The patterns describe resonant modes that are a function of the board geometry, material parameters, port location, and frequency. For a pair of rectangular plates, the resonant frequencies of the modes are given by [86]:

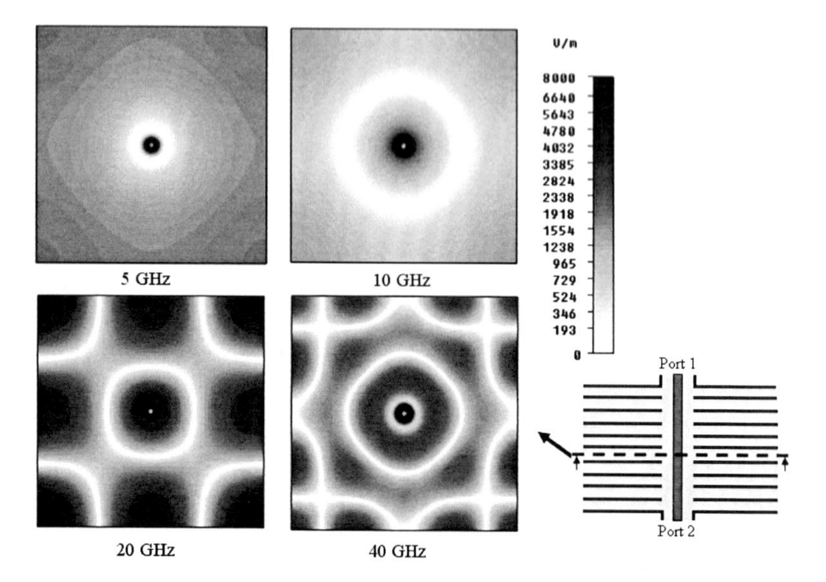

Figure 3.5 Magnitude of the electric field inside the fifth cavity of the structure in Figure 3.3, for zero degree phase and assuming ideal open boundaries (PMC) at the board edges. The plots show the field patterns in presence of reflective boundaries at different frequencies. The excitation is applied to port 1 and the fields are plotted on a logarithmic scale.

$$f_c = \frac{1}{2\sqrt{\mu\varepsilon}}\sqrt{\left(\frac{m}{a}\right)^2 + \left(\frac{n}{b}\right)^2} \quad \text{with} \quad m,n = 0, 1, 2, 3,... \tag{3.6}$$

The parallel-plate impedance (Z^{pp}) for the via segment inside a cavity of the structure of Figure 3.3 is included in Figure 3.6(a). Due to the via position, the first resonance in the PMC case occurs at 15.1 GHz and it is related to the (0,2) and (2,0) modes. Other plate resonances are associated with modes denoted by the integer pairs specified in the Figure 3.6(a). The small plane size locates the resonances at relatively high frequencies. The magnitude of the impedance parameters in Figure 3.6 and the field plots in Figure 3.5 indicate that the parallel resonances are only observable beyond 10 GHz in this example. The simulation of the multilayer via takes into account ten stacked cavities, the near fields in the antipad region, and the ports, which make the analytical estimation of the response more difficult. Nevertheless, the notches on the transfer impedance (Z_{12}, Figure 3.6(b)) can be related to the resonances of the single cavity impedance (Figure 3.6(a)). For absorbing boundaries, which can also be visualized as a frequency-dependent matched load place at finite board edges [85], the impedance of a single cavity increases with the frequency. As expected, the impedance shows a smooth profile in absence of reflections from the board edges.

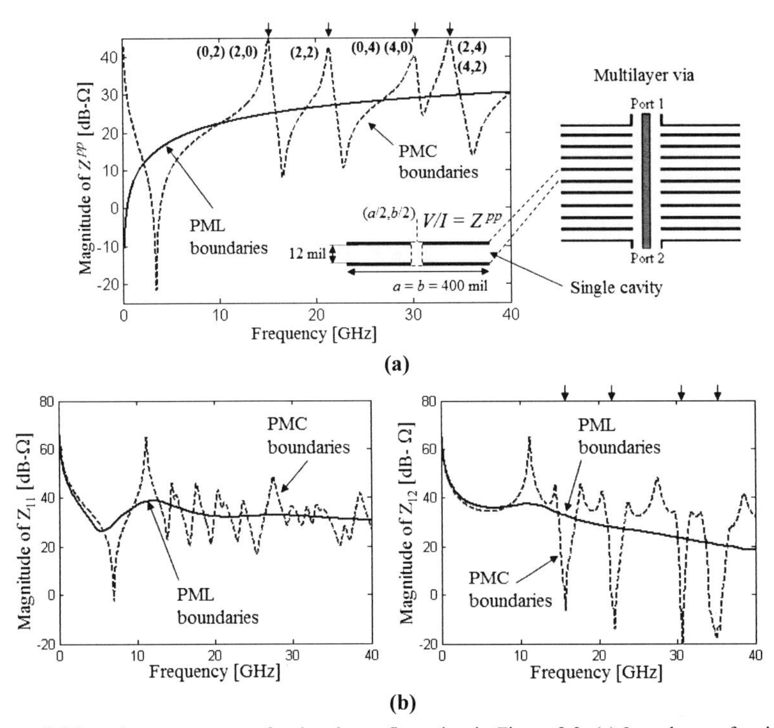

Figure 3.6 Impedance parameters for the via configuration in Figure 3.3. (a) Impedance of a single cavity, called the parallel-plate impedance Z^{pp}. (b) Simulated input and transfer impedances for the multilayer via as a function of the frequency, for PMC (open) and PML (absorbing) boundaries at the board edges.

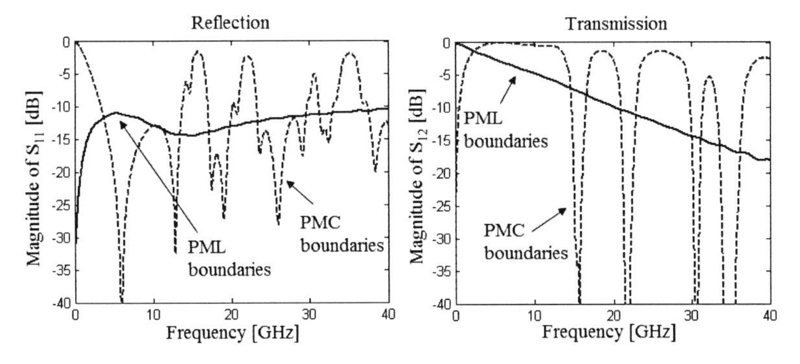

Figure 3.7 Simulated reflection and transmission for the multilayer via in Figure 3.3, using PMC (open) and PML (absorbing) boundaries at the board edges.

The magnitudes of the S-parameters in Figure 3.7, normalized to 50 Ω, show the reflection and transmission between the defined ports at top and bottom sides of the via for the cases of absorbing (PML) and open reflective boundaries (PMC). Absorbing boundaries lead to a higher loss since the fields are only traveling outwards. The field decay is less prominent for open boundaries because of the superposition of incident and reflected waves that results in standing wave patterns. The absence of a DC path for open boundaries is observable in the S-parameters. The parallel-plate impedance concept and methods for its analytical computation are discussed in the next chapter.

3.3. Effect of Ground Vias (Return Vias)

The coupling between vias and reference planes can result in significant crosstalk among interconnects. In addition, the energy coupled into cavity modes affects the transmission of signals over vias. Return vias, usually denoted as ground vias, can help to minimize this coupling by providing a return path for a portion of the via currents flowing on reference planes. Return vias are usually more effective if a large enough number of them are placed in close proximity to signal vias [87].

A simple experiment with ground vias, based on the structure of Figure 3.3, is presented in order to evaluate the effect on the transmission of a signal. A different number of return vias shorting all the reference planes was placed at a radial distance of 40 mils from the centrally located signal via. Figure 3.8 shows the configurations and magnitudes of the complex amplitude of the electric field inside the fifth cavity at 16 GHz. One ground via deforms the field pattern and leads to a considerable amount of energy coupled into the plates. The transmission (S_{12}) in Figure 3.9 indicates that for this case a deep resonance occurs at about 16 GHz. With two, four and six ground vias the transmission is successively improved and the magnitude of the parallel-plate modes tends to be reduced. Ground vias in close proximity can improve the return path and also dampen and shift the plane resonances towards higher frequencies (Figure 3.9). The currents through return vias can be visualized as image sources that create angular modes. These modes tend to cancel the fields in the region around the ground via. Note that several ground vias placed near to a signal one practically replicate a coaxial transmission line [88].

Not only the number but also the location of ground vias is important, since they can also insert resonances in the frequency response [87]. Several ground vias in close proximity allow a more effective confinement of the fields around a signal via. Close placement makes broadband phase cancellation possible. In contrast, a large separation means that phase coherence can only be maintained over small angles and frequency spreads.

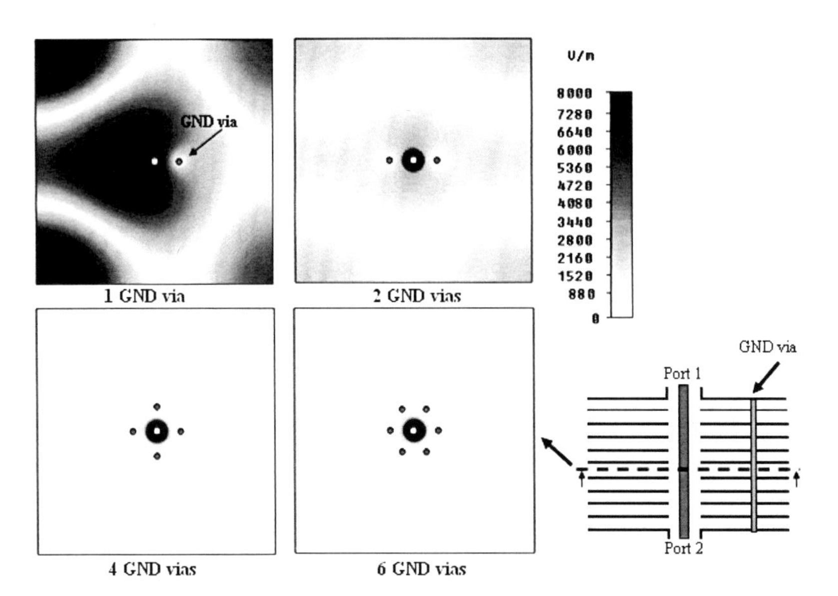

Figure 3.8 Magnitude of the electric field complex amplitude inside the fifth cavity at 16 GHz, considering a different number of ground vias and assuming ideal open boundaries (PMC). The structure is equivalent to the one described in Figure 3.3, but with additional ground vias (5 mil radius, 40 mil pitch). The excitation is applied to port 1 and the plots use a linear scale.

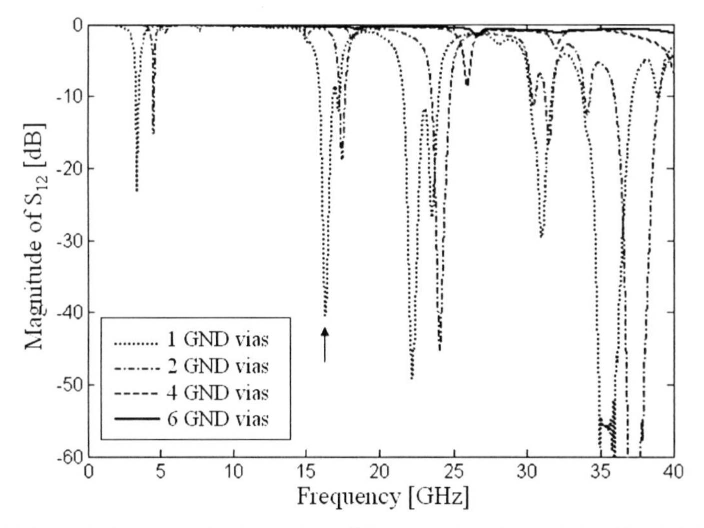

Figure 3.9 Transmission over a signal via using a different number of ground vias. The pitch between the signal and the ground vias was 40 mil, using the constellations depicted in Figure 3.8 and the stackup in Figure 3.3.

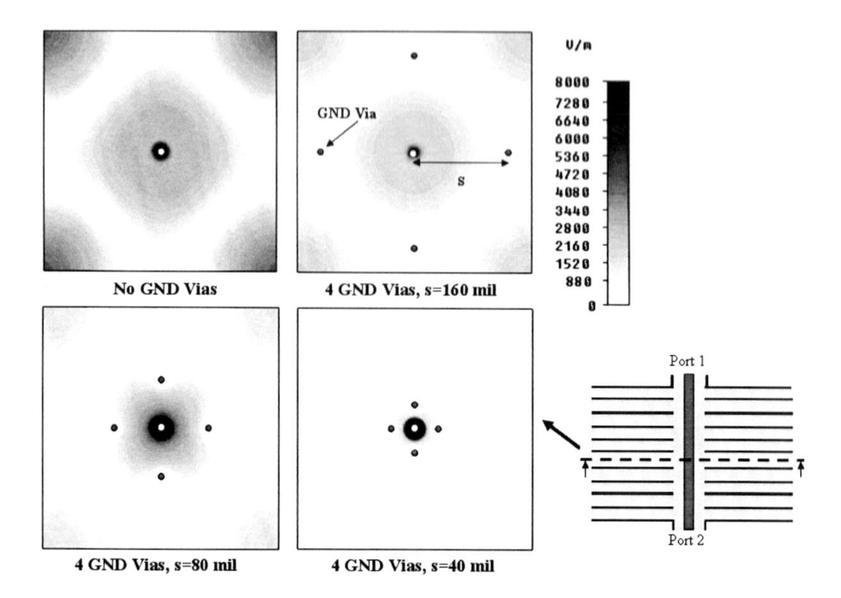

Figure 3.10 Magnitude of the electric field complex magnitude inside the fifth cavity at 16 GHz, considering one signal via and four ground vias placed at different radial separations, and assuming ideal open boundaries (PMC). The structure is equivalent to the one described in Figure 3.3, but with additional ground vias (5 mil radius). The plots use a linear scale.

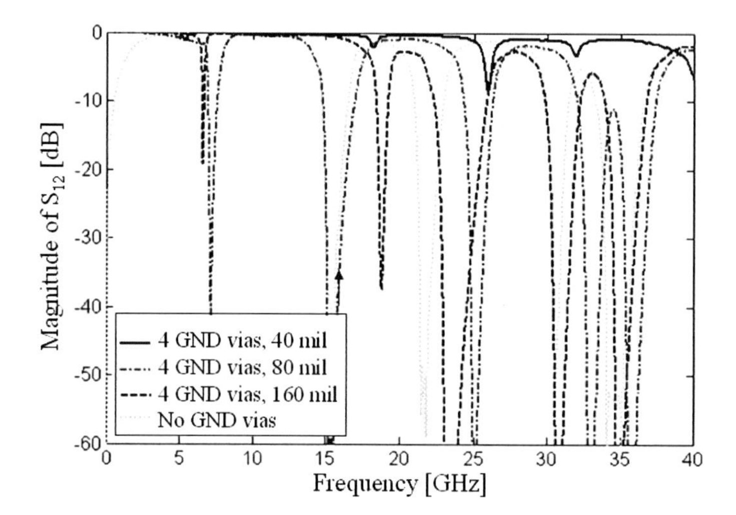

Figure 3.11 Transmission over a signal via using four ground vias placed at different radial distances, distributed according to Figure 3.10 and using the stackup in Figure 3.3.

An example with four ground vias placed at different distances is discussed for the same configuration, as detailed in Figure 3.10. At 16 GHz, as the ground elements approach the central via, the energy spread inside the cavity is reduced. In Figure 3.11 it is observed that, in general, the transmission is improved as the separation s is reduced. It can also be shown that larger ground vias can provide a better return path; however high density boards usually require the utilization of smaller interconnect elements.

3.4. Via Crosstalk

The excitation of parallel-plate modes may induce a considerable amount of crosstalk among vias. The interaction between elements depends, as shown before, on how effectively the energy is coupled into the cavities, the position of the vias, and their environment. The utilization of ground vias is an important tool for mitigating the crosstalk [88].

Figure 3.12 depicts a modified example, based again on the configuration in Figure 3.3, with two signal vias and four ports defined at the via ends. The electric field distribution shows the presence of plate modes and the field coupling to the via on the right side. This vector distribution was plotted on a diagonal cut for the case without

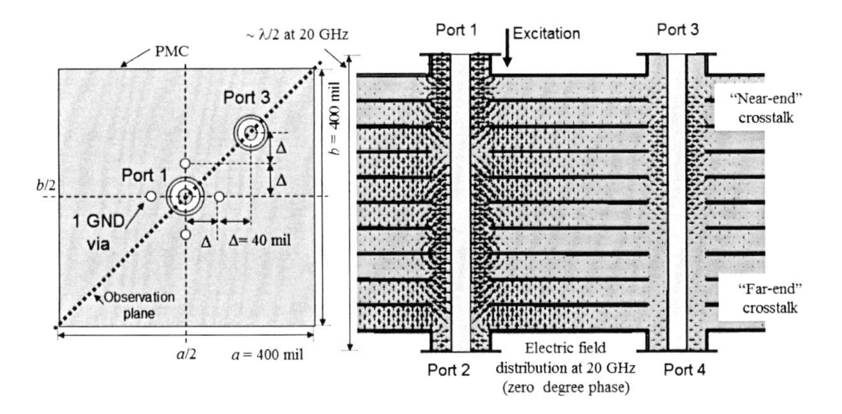

Figure 3.12 FIT simulation [83] of crosstalk among two vias crossing a 10-cavity stackup, based on the geometry defined in Figure 3.3. The electric field distribution was plotted along the diagonal observation plane at 20 GHz (zero degree phase), for the configuration without ground vias. The excitation ($1-(W)^{0.5}$ Gaussian pulse, $f_{max} = 40$ GHz) was applied to port 1.

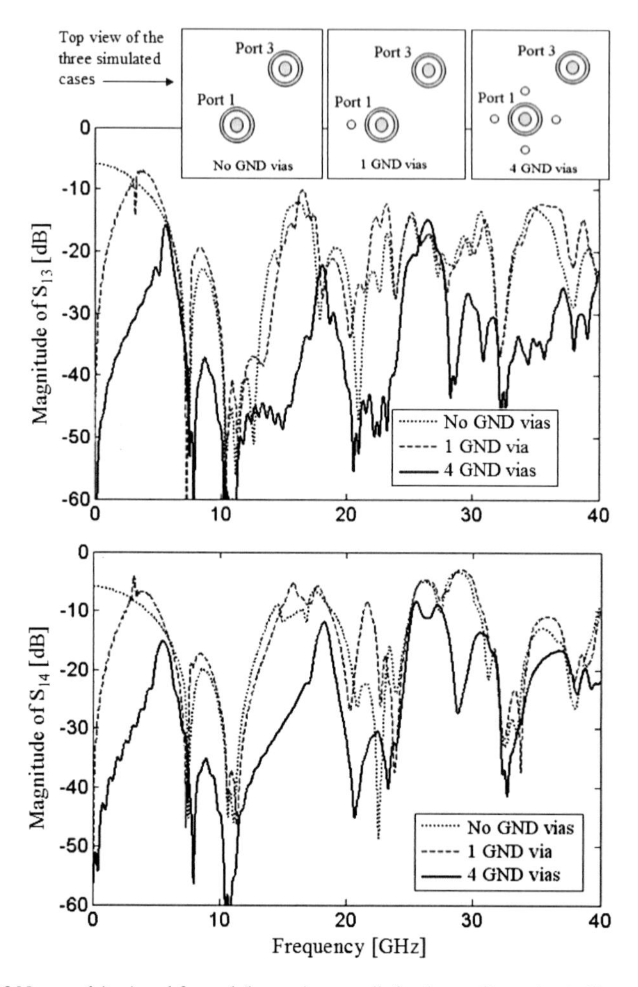

Figure 3.13 Near-end (top) and far-end (bottom) crosstalk for the configuration in Figure 3.12.

ground vias. The simulated crosstalk in terms of *S*-parameters is included in Figure 3.13, where the worst coupling predicts a level over -20 dB. Three cases were simulated: no ground vias, one return via located on the left side of the board, and four ground vias surrounding the central via. The simulation indicates that one ground via can help to reduce the crosstalk mainly at lower frequencies, but only marginally in the GHz range. Four return vias prove to be more effective over a broader bandwidth.

As a first-order approximation, the coupling through Z^{pp} is mainly inductive [49]. At higher frequencies the coupling mechanisms become more complicated and require more rigorous modeling.

3.5. Transmission Line Parameters of Via Interconnects

The proper design of vias is a challenging task given the complex coupling mechanisms among many-element configurations and the multiple geometrical parameters that must be considered. Optimal design solutions should deal not only with electrical performance, but also achievable interconnect density and cost.

In order to study the role of different via parameters, the basic configuration in Figure 3.3 will be examined again. Since the structure is symmetric ($S_{11}=S_{22}$) and reciprocal ($S_{12}=S_{21}$), an equivalent characteristic impedance and propagation constant can be extracted [89],[90]. Transforming the simulated *S*-parameters to an *ABCD* matrix and assuming a reference system impedance of 50 Ω (see Appendix A.2), the transmission line parameters can be readily identified [90]

$$\begin{bmatrix} A & B \\ C & D \end{bmatrix} = \begin{bmatrix} \cosh(\gamma l) & Z^{v0} \cdot \sinh(\gamma l) \\ \dfrac{\sinh(\gamma l)}{Z^{v0}} & \cosh(\gamma l) \end{bmatrix}, \tag{3.7}$$

with γ the via propagation constant and Z^{v0} the via impedance, which can be found as

$$Z^{v0} = \sqrt{B / C}\,, \tag{3.8}$$

$$\gamma = \alpha + j\beta = \cosh^{-1}\left(A\right) / l\,, \tag{3.9}$$

where l denotes the via length, α the attenuation constant and β the angular phase constant. The propagation velocity of the via can be computed from

$$v_p = \frac{\omega}{\beta} = \frac{\omega}{imag(\gamma)}. \tag{3.10}$$

To get β, the phase of $\beta \cdot l$ must be "unwrapped". This means that it must be transformed from the cyclical phase $[-\pi,\pi]$ to its real value in radians $(\theta+2\pi n)$ [90]. As described in the literature, the transmission line parameters can also be obtained directly from the *S*-parameters [89].

The via impedances as a function of the frequency for the case in Figure 3.3, defining absorbing and open boundaries, are compared in Figure 3.14. The plot also includes the

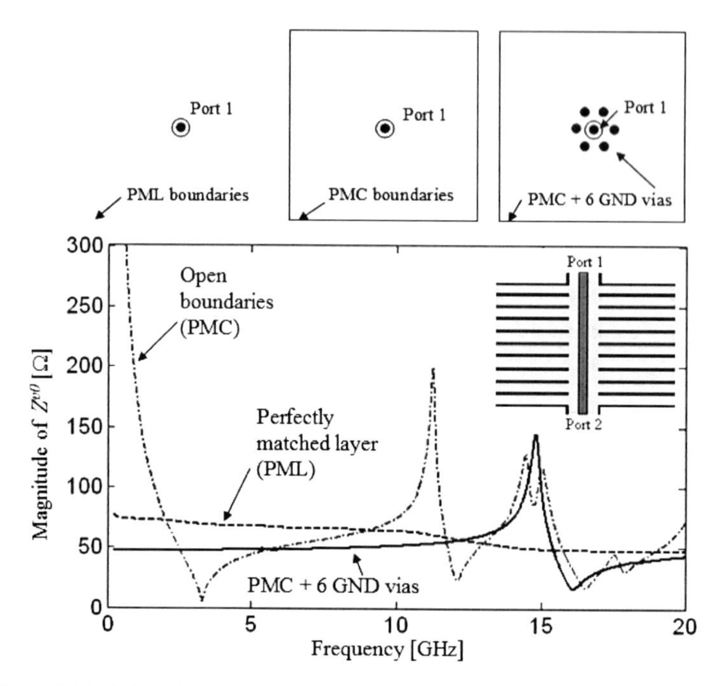

Figure 3.14 Via impedance for different boundary conditions at the board edges, and in presence of ground vias. The structure is shown in Figure 3.3 and the ground via locations in Figure 3.8 (5-mil via radius, 40 mil pitch between signal and ground vias). The reference impedance is 50 Ω.

case with six ground vias and PMC boundaries, as defined for the example in Figure 3.8. The results are shown in the frequency range from 0.2 up to 20 GHz, where clear trends are observable. The accuracy limit of the FIT simulation makes it difficult to extract the parameters at very low frequencies. The ideal case of a single via with absorbing boundaries shows a smooth decaying variation of the impedance with respect to the frequency. In contrast, with open boundaries the broadband impedance control in presence of plate resonances is not possible. By adding six ground vias to the open boundary case, the effect of plate resonances can be mitigated and a very good impedance control can be achieved up to 12 GHz. In fact, an approach to design impedance controlled vias is to emulate a coaxial configuration, by placing return vias very close to the antipad rim and sizing the coaxial region to match the target impedance [88],[91]. The effect of ground vias is however band limited and plate resonances or inter via resonances may still appear at higher frequencies, as shows Figure 3.14 at about 15 GHz.

Figure 3.15 compares the attenuation constant and the normalized propagation velocity for the cases of infinite planes and six ground vias. The parallel plates lead to a

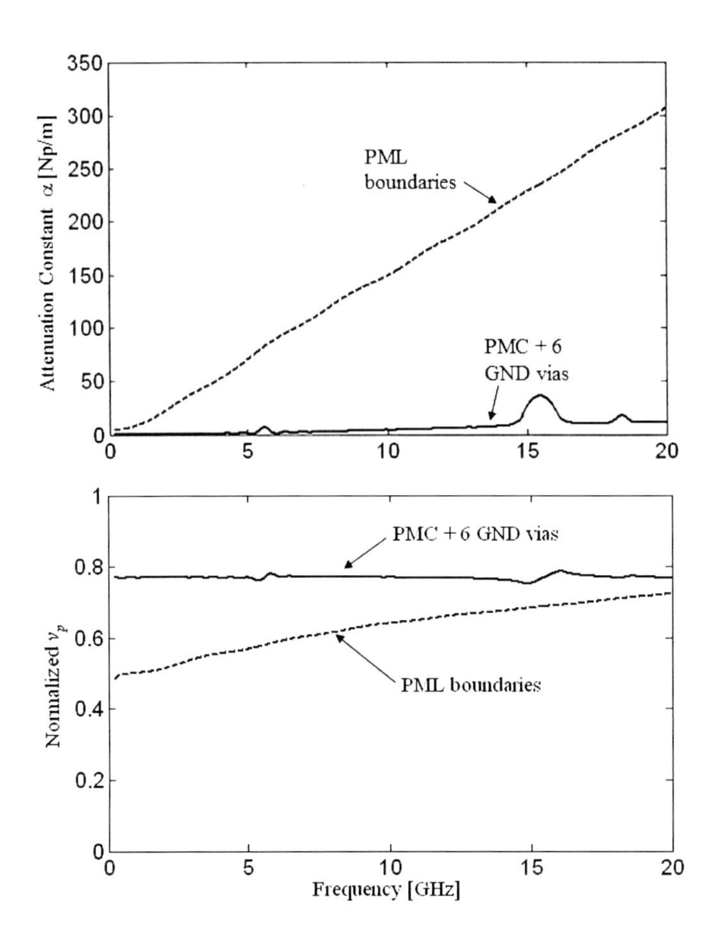

Figure 3.15 Attenuation constant and normalized propagation velocity for the cases of absorbing and open boundaries with six ground vias (cases in Figure 3.14). The propagation velocity is normalized to $c_0/(3.8)^{0.5}$.

"slower" wave propagation in the examined bandwidth, which is far below the plane wave propagation in the dielectric medium. Below 20 GHz, the propagation velocity for the case with six ground vias is higher when compared to the case of absorbing boundaries without ground vias, and thus the via appears electrically shorter [13]. Ground vias also help to flatten the frequency dependence, which leads to less dispersion [89]. The attenuation constant is much larger for the case with infinite planes since the return path provided by ground vias confines the fields around the signal via more efficiently, at least within the discussed bandwidth. In presence of severe plate

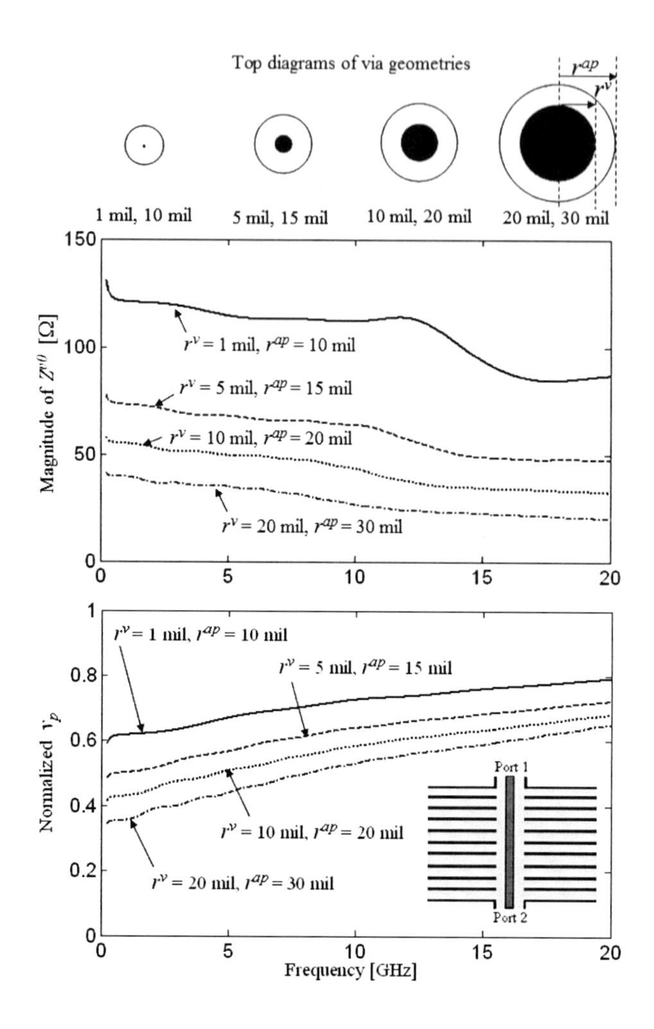

Figure 3.16 Impedance and propagation velocity of a single via for different sizes, defined by the via radius r^v and antipad radius r^{ap}, assuming absorbing boundaries. The basic configuration is shown in Figure 3.3. The reference impedance is 50 Ω and the propagation velocity is normalized to $c_0/(3.8)^{0.5}$.

resonances the estimation of the phase constant and the propagation velocity become difficult due to the rapid magnitude and phase variations at resonant frequencies.

The effect of the via size for a single element and assuming infinite planes is shown in Figure 3.16. The via radius has been varied from 1 to 20 mil, keeping a constant

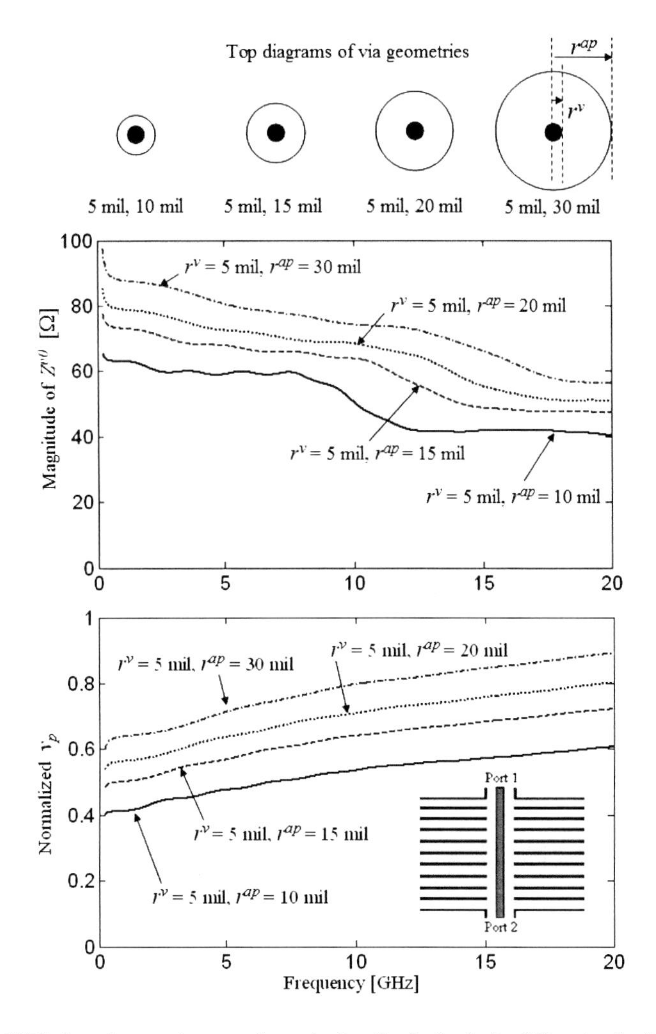

Figure 3.17 Via impedance and propagation velocity of a single via for different antipad sizes (r^{ap}) and a via radius (r^v) of 5 mil, assuming absorbing boundaries. The basic configuration is shown in Figure 3.3. The reference impedance is 50 Ω and the propagation velocity is normalized to $c_0/(3.8)^{0.5}$.

separation between the via barrel and the antipad rim of 10 mil. The simulation results show that both the via impedance and propagation velocity tend to increase as the via radius decreases. For a higher interconnect density and low loss, small vias are more adequate, however their impedance control may turn to be more difficult due to more

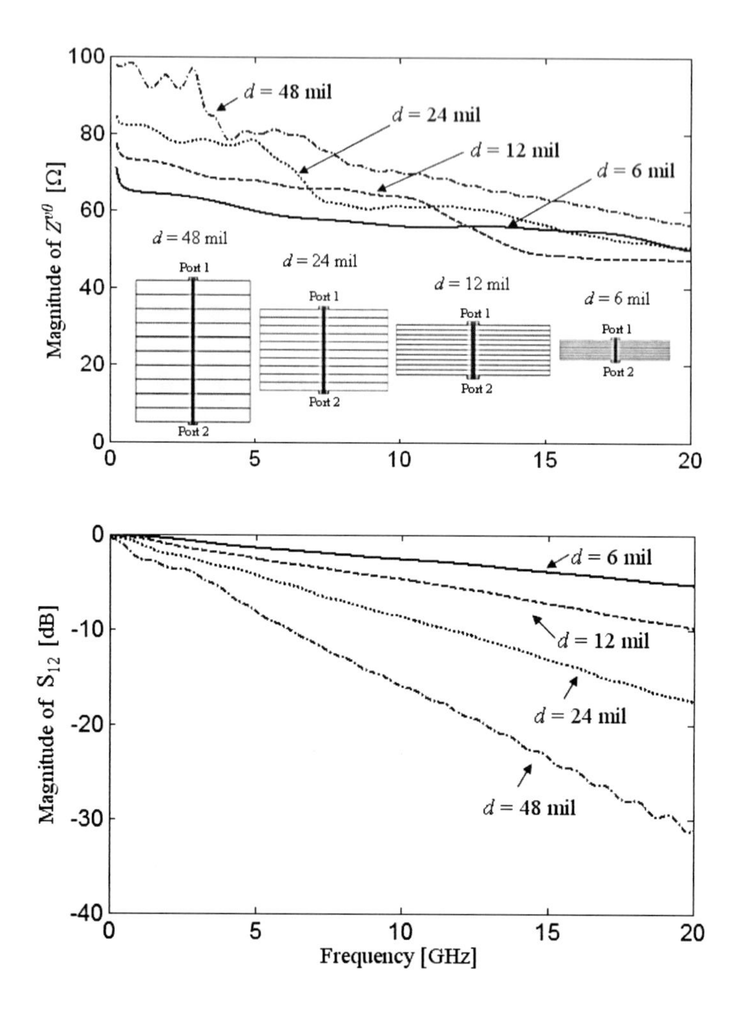

Figure 3.18 Via impedance and transmission for different cavity thicknesses *d*, a via radius of 5 mil and an antipad radius of 15 mil, assuming infinite planes. The basic configuration is shown in Figure 3.3. The reference impedance is 50 Ω.

pronounced frequency dependencies and larger impedance values. Minimum feature sizes are also restricted by technology and fabrication processes.

Figure 3.17 displays the results for different antipad sizes while keeping a constant via radius of 5 mil. A larger antipad is associated with a smaller capacitance between the via barrel and the reference planes, which leads to an increment of the via impedance and the propagation velocity. The size of the via barrel and the antipad are in practice restricted by the geometrical form factor of the interconnects being routed. Larger antipads may reduce the coupling between vias and planes. This is however incompatible with high interconnect densities and the perforations in reference planes are detrimental for traces, which must cross longer regions without a proper return path and may also suffer from inter-cavity crosstalk. Note that the effect of via pads is analogous to the problem of antipad size. The via pads basically increase the via capacitance and therefore their presence tends to decrease the impedance and to limit the propagation velocity.

A last experiment considers the role of the plane separation. The impedance and the transmission for different cavity thicknesses d are given in Figure 3.18. As will be shown in the next chapter, the impedance between two plates increases with d. A small plate separation is usually desirable due to the higher inter-plate capacitance for PDN decoupling at low frequencies, the mitigation of parallel-plate modes inside the cavity, and the lower radiated emissions through board edges. A shorter via, obviously, is associated with a lower material loss. Larger plane separations tend to increase the via impedance, at least for the lower frequency band, and it leads to a poorer broadband impedance control. Nevertheless, the target impedance for traces also imposes a limitation on the plane separation.

The cases that have been discussed in this section addressed significant trends for via design, however the via geometry alone does not determine the impedance and propagation velocity. The return path is also important which means that the via environment should not be overlooked.

3.6. The Via Stub Effect

In the transition from a via to a stripline, the remaining via section below the trace –or above depending on port location, called *via stub*, may introduce undesirable resonances. Stub vias behave like open ended transmission lines that can act as notch filters [81],[92]. For a given via environment, the notch frequency is inversely proportional to the stub length. Longer stubs shift resonances towards lower frequencies and therefore reduce the usable bandwidth of the link.

Figure 3.19 shows a simple configuration that was simulated with the FIT method in order to illustrate this effect. A via crossing a different number of cavities is connected to a stripline inside the first cavity. One waveguide port is defined on top of the via and the second at the end of the stripline. Other geometrical and material parameters correspond to the ones defined for the reference example in Figure 3.3. The stub segment is identified with the length ds_i for the considered cases. The transmission in terms of S-parameters is plotted in Figure 3.20(a). The first order $\lambda/4$ stub notch and

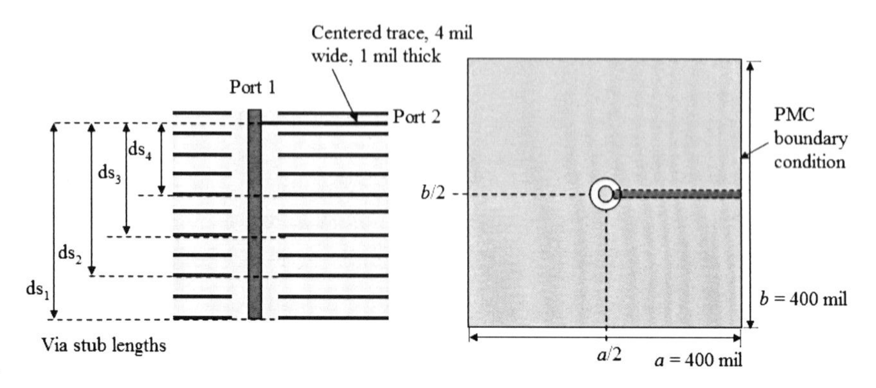

Figure 3.19 Description of the structure for simulation of the via stub effect. Via sizes and material parameters correspond to Figure 3.3. The simulations were done with the finite integration method [83].

other resonances due to cavity interaction are observable in the response. As expected, longer stub lengths lead to notches located at lower frequencies. Although the link might be still utilized for narrow band systems, in general, the transmission of broadband signals can become very difficult, if possible at all. Multiple via stubs may introduce additional inter-via resonances that are detrimental for the link performance as well [81].

It is usually desirable to reduce or eliminate the via stub segments. This can be achieved by placing the traces at deeper levels, by mechanical removal of stub segments (e.g. back-drilling) or by using buried/blind via construction.

The notches can be shifted to higher frequencies by reducing the via-to-plane capacitance, for instance, by eliminating pads on unconnected signal levels and by optimizing the antipad size. The electrical length of the stub is however not solely determined by the via geometry, the return path also plays an important role here. The utilization of ground vias in close proximity can contribute to mitigate the stub effect. Figure 3.20(b) illustrates the effect of a smaller antipad and ground vias on shifting the first stub resonance. A smaller antipad, which translates into a larger via-to-plane capacitance, reduces the effective propagation velocity of waves along the via and the notches appear at lower frequencies. With the utilization of ground vias, the stub notch can be pushed towards higher frequencies. The ground vias tend to increase the wave propagation velocity and hence the stub length appears electrically shorter than when ground vias are absent or placed at larger radial distances.

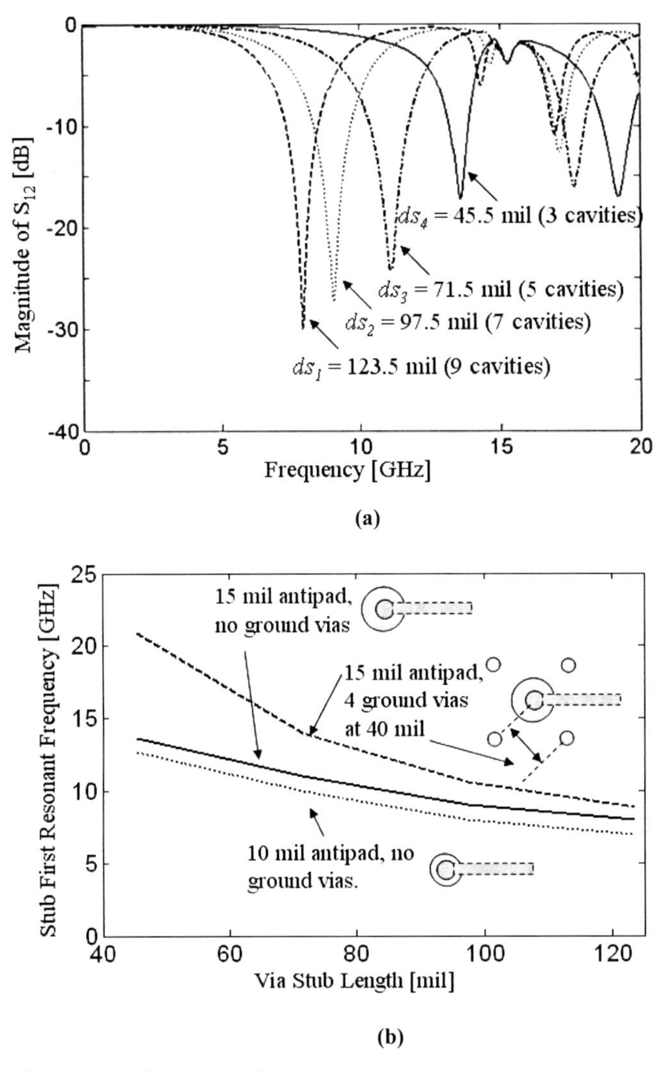

Figure 3.20 Simulation of the via stub effect. (a) Transmission for the structure in Figure 3.19 for different stub lengths. The longer the via stub length the lower the frequency associated with the first notch. (b) The stub first resonance is plotted as a function of the via stub length. Three cases are considered: the reference structure in Figure 3.19, the same configuration reducing the antipad radius from 15 mil to 10 mil, and the original geometry with four ground vias with a 40-mil pitch (signal-to-ground).

4. Development of Semi-Analytical Via and Trace Models

The models developed in this work and the framework for high-frequency simulation of multilayer substrates are presented in this chapter. First, the modeling approach and its advantages are discussed. Then, the physics-based via model and its formulation using microwave network parameters are described. The computation of the via-model building blocks is addressed by introducing analytical methods to compute the parallel-plate impedance and alternatives to calculate or extract the via-to-plane capacitances. The extension of the models to handle the via-stripline transition by applying modal decomposition is explained as well. The general method and the framework for simulation of multilayer substrates are presented in the last section.

4.1. Modeling Approach

As discussed in Section 2.5, different approaches to describe vias and traces in parallel-plate environments can be found in the literature. A classification for via modeling techniques has been presented by C. Schuster *et al.* in [80]. Quasi-static formulae and numerical methods have been used in the past to describe simplified configurations. These models are instructive to gain insight on the problem, but the resultant equivalent circuits, typically π-type RLC networks, are restricted to simple configurations and limited to work only at relative low frequencies [93].

The high-frequency characterization of multilayer substrates requires more rigorous approaches for simultaneous modeling of complex configurations with vias and traces of the signal network, and reference planes and power/ground vias of the power distribution network. Since the size of the total structure may become comparable or even much larger than the minimum wavelength of interest, it is necessary to consider electromagnetic wave effects. General purpose numerical methods are in principle capable of handling this task. These techniques can provide a good accuracy and flexibility; they are capable of simulating irregular configurations and capturing 3D dynamic field effects that are otherwise difficult to describe. Nevertheless, the setup of a geometrical model is required and the computational burden rapidly grows to prohibitive levels as the geometrical and electrical size, and complexity of the interconnect structure increase.

To find alternatives for efficient simulation of multilayer substrates has been a topic of active research in the last decades, as mentioned in Section 2.5. Customized numerical approaches for analysis of via transitions and power planes have been proposed in the past, many of them exploiting the planar nature of the PDN. Although these techniques require either the setup of large equation systems or the discretization of at least some regions of the computation domain, their numerical efficiency is usually much higher in comparison to general purpose numerical methods. Analytical or semi-analytical formulae are also available. They can provide the highest numerical efficiency but have limitations to model irregular structures.

Customized numerical or analytical methods often require the partitioning of the computational domain under certain assumptions. The hybridization of different methods and mathematical techniques is then applied to solve the different sub-domains and to combine the partial results. This is an attractive approach since the best characteristics of different methods can be exploited.

The models used and developed in this work rely on the hybridization of diverse analytical and numerical techniques in the frequency domain. It has been also denoted in previous publications as "physics-based" (e.g. in [19]) meaning that the selected partitioning approach requires the identification of the main physical mechanisms governing the behavior of the structure being described. The topology of the models –i.e. the relation between their different components– has been analytically derived and formulated at network level, in terms of microwave network parameters (Appendix A.2). This representation allows the extension of the models to an arbitrary number of elements by just changing the dimensions of the matrices. This is advantageous from a system-level perspective for the study of complex configurations and for the automation of the simulation process. The models are denoted as semi-analytical since each of their components can be independently computed by using either numerical or analytical formulations. Their complexity can also be scaled according to the physical structure being described.

In summary, the main advantages of the approach proposed in this work are:

- The models are concise and general enough to handle an arbitrary number of elements. The chosen partitioning approach allows a clear definition of interfaces interconnecting the model building blocks. This makes it possible to develop a better insight into the physics of the problem, to evaluate the contribution of the different model components, and to increase the numerical efficiency.

- Analytical formulae to compute most model components are available. They can further enhance the numerical efficiency and enable the simulation of complex structures.
- The models can be fully parameterized –and given the achievable efficiency– they are appropriate for fast prototyping, trade-off, and optimization studies.

4.2. Via Model and Its Formulation using Microwave Network Parameters

The physics-based via model utilized as starting point for this work had already been proposed and validated in [71],[80],[94]-[96]. The model is inspired by previous solutions found in the literature [70],[93],[97]-[99], which make use of a layered partitioning approach (Figure 4.1). Its basic cell corresponds to the via segment crossing a cavity enclosed by two adjacent reference planes. The network representation introduced here [19] consists of a π-connection of different elements, associated with the two main physical mechanisms describing the via transitions:

- The *parallel-plate impedance* Z^{pp}. It models the return path for vias by describing the interaction between the via transitions and the reference planes in terms of propagating cavity modes (see Section 4.3). As a first order approximation for low frequencies, this impedance can be approximated by an equivalent inductance [100]-[101], connected in series with the static parallel-plane capacitance for the case of finite planes.
- The *via barrel-to-plane capacitance* C^v. It approximates the near field around via transitions. This capacitance accounts for the contribution of the evanescent modes in the antipad region, assuming that there is no near-field coupling among vias (see Section 4.4).

In this formulation it is assumed that there is no field penetration through the reference planes and no interactions at the board edges. The vias offer thus the only coupling path among the different cavities of multilayer structures. Furthermore, the vias are assumed to be small in comparison to the wavelength; it is supposed that the via currents are uniform and consequently their excitation fields show cylindrical symmetry.

For electrically small ports, Z^{pp} is interpreted as the plane impedance seen at the via locations, approximated as the ratio of voltage to current at each via segment (V_i / I_i). Also, it is assumed that the cavity thickness is small in comparison to the minimum wavelength of interest and therefore the fields inside the cavity do not vary in the direction perpendicular to the plates (z-direction). The parallel-plate impedance formulae, discussed in the next section, are used to construct the two-dimensional

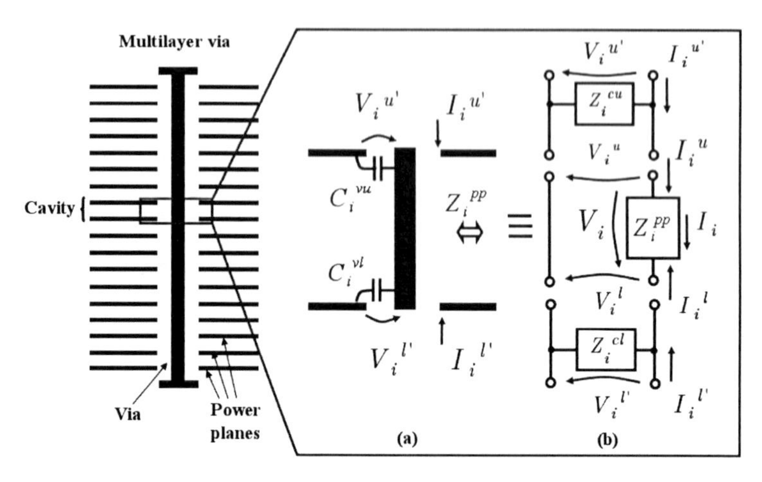

Figure 4.1 Description of the via model for a single via. (a) *i-th* via transition basic cell. (b) Its equivalent network representation. Z_i^{pp} denotes the parallel-plate impedance and Z_i^c the impedance associated with the via capacitances $(1/j\omega \cdot C_i^v)$. Superscripts *u* and *l* refer to the *upper* or *lower* cavity sides, respectively.

impedance matrix per frequency point

$$\overline{V} = \overline{\overline{Z}}^{pp} \cdot \overline{I}, \tag{4.1}$$

where the vectors $\overline{V}$ and $\overline{I}$ contain the voltages and currents defined for each via segment, respectively. The via barrel radius is used to define the port size in the computation of Z^{pp}. Since it is convenient to provide top and bottom connection nodes for the vias, in order to interconnect multiple cavities, the following voltage and current relations are applied on each port i [19]

$$V_i = V_i^u - V_i^l, \qquad I_i = I_i^u = -I_i^l. \tag{4.2}$$

The superscript u stands for the upper side and the superscript l for the lower side of the cavities. The resulting matrix in terms of microwave network parameters contains the explicit definition of ports on both sides, for instance, as Y-parameters

$$\begin{bmatrix} \overline{I}^u \\ \overline{I}^l \end{bmatrix} = \underbrace{\begin{bmatrix} \overline{\overline{Y}}^{pp} & -\overline{\overline{Y}}^{pp} \\ -\overline{\overline{Y}}^{pp} & \overline{\overline{Y}}^{pp} \end{bmatrix}}_{\overline{\overline{Y}}^{v}} \cdot \begin{bmatrix} \overline{V}^u \\ \overline{V}^l \end{bmatrix}, \tag{4.3}$$

with $\overline{\overline{Y}}^{pp} = \overline{\overline{Z}}^{pp^{-1}}$. Appendix A.3 provides the mathematical procedure to expand the matrices. The Y-matrix can be easily extended to consider the via capacitances by writing the π-network as block matrices according to [19]

$$\begin{bmatrix} \overline{I}^{u'} \\ \overline{I}^{l'} \end{bmatrix} = \left[\underbrace{\begin{bmatrix} \overline{\overline{Y}}^{cu} & 0 \\ 0 & \overline{\overline{Y}}^{cl} \end{bmatrix}}_{\overline{\overline{Y}}^{c}} + \underbrace{\begin{bmatrix} \overline{\overline{Y}}^{pp} & -\overline{\overline{Y}}^{pp} \\ -\overline{\overline{Y}}^{pp} & \overline{\overline{Y}}^{pp} \end{bmatrix}}_{\overline{\overline{Y}}^{v}} \right] \cdot \begin{bmatrix} \overline{V}^{u'} \\ \overline{V}^{l'} \end{bmatrix}. \tag{4.4}$$

The capacitive elements are arranged in a diagonal matrix, in which each entry is defined as $Y_i^c = 1/Z_i^c = j\omega \cdot C_i^v$ for the top and bottom sides, respectively. The vectors of size n, $\overline{V}^{u'}$ and $\overline{I}^{u'}$, denote the via voltages and currents defined on the upper cavity side, and the vectors $\overline{V}^{l'}$ and $\overline{I}^{l'}$, on the lower side. Note that the formulation is general for an arbitrary number of via transitions.

4.3. Analytical Computation of the Parallel-Plate Impedance

The parallel-plate impedance has shown to be a useful concept to model power/ground planes of PCBs and packages in the GHz range. A wide variety of techniques have been used in the past such as numerical methods [54]-[55], the contour integral method [61], scattering methods [65], as well as analytical formulations [70]. Although the analytical solutions are restricted to relatively simple board shapes, they constitute an attractive alternative given their high computational efficiency, which allows rapid handling of relatively complex interconnect arrays.

In this section, two different analytical approaches to calculate the parallel-plate impedance for rectangular planes are going to be discussed: the *cavity resonator model* (CRM) [60],[69] and the *radial waveguide method* (RW) [85],[102]. The first technique considers a plate pair subject to a given boundary condition at the periphery (Figure 4.2(a)). The impedance at port locations is obtained by expressing the 2D Helmholtz equation as an eigenmode expansion of the Green's function. The second technique takes advantage of the cylindrical symmetry of ports on infinite planes to describe the impedance in terms of TEM waves propagating radially (Figure 4.2(b)). The two methods have been successfully applied to model problems involving PCB and package structures; for instance, in [103] the cavity model is used, and in [70] the radial waveguide approach. Comparative analyses, convergence properties, guidelines for the utilization of one or the other approach, and the hybridization of the methods for an efficient computation of Z^{pp} are going to be discussed in the following sections.

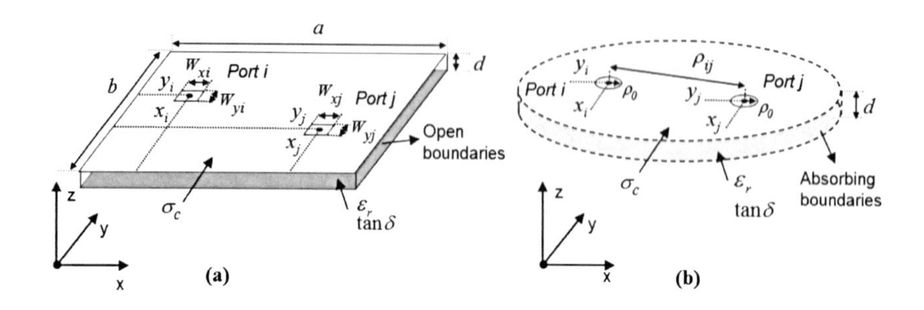

Figure 4.2 Diagrams of the parallel-plate impedance Z^{pp} calculation domains. (a) Cavity resonator model. (b) Radial waveguide model for infinite plates.

4.3.1. The Cavity Resonator Method

Assuming that the plate separation is small in comparison to the wavelength, the cavity formed between two power planes can be described by the 2D Helmholtz equation under given boundary conditions on the sidewalls of the cavity [60],[69]. According to the drawing in Figure 4.2(a), the impedance between two arbitrary ports on rectangular plates can be expressed as a function of the frequency ω as

$$Z_{ij}^{CRM}(\omega)=\frac{j\omega\mu d}{ab}\sum_{m=0}^{\infty}\sum_{n=0}^{\infty}c_m^2c_n^2\cdot\frac{f_{B.C.}\cdot f_P}{k_m^{\ 2}+k_n^{\ 2}-\underline{k}^2},\tag{4.5}$$

where the superscript *CRM* stands for *cavity resonator model*, $c_m,c_n=1$ for $m,n=0$, and $c_m,c_n=\sqrt{2}$ for $m,n\neq 0$, $k_m=m\pi/a$, and $k_n=n\pi/b$. The function $f_{B.C.}$ reflects the boundary condition defined at the cavity sidewalls. For rectangular plates and open boundaries, approximated as a perfect magnetic conductor wall (PMC), the function can be written as [60]

$$f_{\text{B.C.}(PMC)}=\cos\left(k_m x_i\right)\cdot\cos\left(k_n y_i\right)\cdot\cos\left(k_m x_j\right)\cdot\cos\left(k_n y_j\right).\tag{4.6}$$

The port center locations are indicated by x_i, x_j and y_i, y_j. For perfect electric conductor (PEC) boundaries, which represent a cavity enclosed by metallic walls, the boundary function becomes [80]

$$f_{\text{B.C.}(PEC)}=\sin\left(k_m x_i\right)\cdot\sin\left(k_n y_i\right)\cdot\sin\left(k_m x_j\right)\cdot\sin\left(k_n y_j\right).\tag{4.7}$$

The function f_p results from the finite port size, denoted by the port side lengths W_{xi}, W_{yi}, W_{xj} and W_{yj}. For rectangular ports the function is given by

$$f_P = \mathrm{sinc}\left(\frac{k_m W_{xi}}{2}\right)\cdot\mathrm{sinc}\left(\frac{k_n W_{yi}}{2}\right)\cdot\mathrm{sinc}\left(\frac{k_m W_{xj}}{2}\right)\cdot\mathrm{sinc}\left(\frac{k_n W_{yj}}{2}\right), \tag{4.8}$$

with $\mathrm{sinc}(x) = \sin(x)/x$.

Dielectric and electric conductor losses are taken into account by the complex wave number $\underline{k}$, according to [60]

$$\underline{k} = \omega\sqrt{\mu_d \varepsilon_d}\,(1 - j(\tan\delta + t_s/d)/2), \tag{4.9}$$

where $\tan\delta$ denotes the dielectric loss tangent, and $t_s = \sqrt{2/\omega\mu_c\sigma_c}$ the conductor skin depth. The formula assumes that the loss is small enough to not disturb the field distribution (perturbation method) and a well developed skin effect. Therefore, the loss model is not accurate for very low frequencies [104]. The subscripts d and c related to Eq.(4.9) stand for *dielectric* and *conductor*, respectively.

The evaluation of Eq. (4.5) becomes time consuming when many modes are considered. Different alternatives can be found in the literature to reduce the numerical burden [105]-[108]. For instance, as detailed in [105], the expression in Eq. (4.5) can be reduced to a single summation by using a Fourier series formula. Collapsing the port in x-direction and for PMC boundaries, it is found that the impedance takes the form

$$\begin{aligned} Z_{ij}^{CRM}(\omega) = \frac{\omega\mu a d}{j2b}\sum_{n=0}^{\infty} c_n^2 \cdot \cos\left(k_n y_i\right)\cdot\cos\left(k_n y_j\right) \\ \cdot\ \mathrm{sinc}^2\left(\frac{k_n W_y'}{2}\right)\cdot\left[\frac{\cos\left(\alpha_n x_-\right)+\cos\left(\alpha_n x_+\right)}{\alpha_n\cdot\sin\left(\alpha_n\right)}\right], \end{aligned} \tag{4.10}$$

with $\alpha_n = a\cdot\sqrt{\underline{k}^2 - k_n^2}$, $x_\pm = 1 - |x_i \pm x_j|/a$ and $W_y' = W_x + W_y$ for ports inside the board. Different scenarios and numerical issues related to the evaluation of Eq. (4.10) are discussed in [105],[109].

In this method, a circular port geometry needs to be approximated by an equivalent square-shaped region. Different port definitions have been used in the past, including the assumptions that either the port width $W = W_x = W_y$ in the cavity model corresponds to two times the circular port radius (*i.e.* circular port circumscribed by a square region), that the circular port and the square one have equivalent areas, or that the port sizes provide an equivalent circular perimeter. These three approximations give conversion factors of 2, 1.77 and 1.57 between the circular radius and the port width

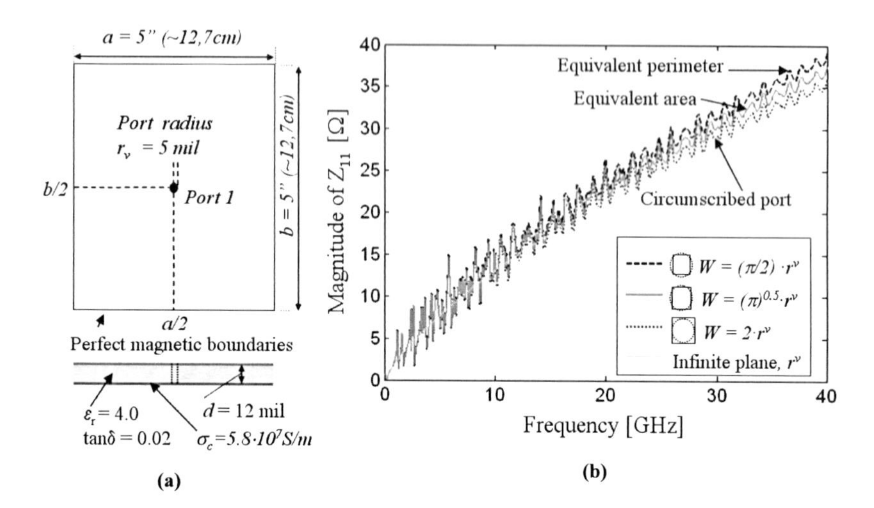

Figure 4.3 Effect of port approximation for circular geometries in the cavity resonator model. (a) One-port case description. (b) Magnitude of the self-impedance for three different definitions of the rectangular port size in the double summation: circular port circumscribed, equivalent circular port area, and equivalent circular port perimeter. The prediction assuming infinite plates is also given, whose formulation assumes circular ports.

W, respectively. If the ports are very small in comparison to the wavelength, these definitions do not play an important role for the computation of the impedance parameters; however they start to have an impact on the results as the physical dimension of ports and the frequency range of interest increase. In particular, this is relevant for closely spaced ports, where the self-impedances represent the worst case. Figure 4.3 provides one example that shows the impact of different port approximations on the self-impedance. Beyond 20 GHz the results of the three alternatives start to deviate noticeably from each other, with differences of over 10% among the self-impedances. The values also start to diverge from the infinite plane solution, which is formulated for circular ports as explained in the next section. A larger impedance value can be associated with a port parasitic inductance that is inversely proportional to the port size [105].

It has been observed that the port definition in the cavity model has a relative weak impact on the correlation studies done for multilayer structures, which are discussed in the next chapters. However, further investigations are required for a better understanding of its role in relation to the via model, the via near fields, and the limitations of the assumption of electrically small ports.

4.3.2. The Radial Waveguide Method

In absence of reflective boundaries, the cylindrical symmetry of the problem can be exploited to compute Z^{pp} (Figure 4.2(b)). As in the cavity model, since the plate separation is small, the analysis is reduced to two dimensions. According to [85],[102], the impedance at every port location can be computed by

$$Z_{ij}^{RW}(\omega) = \frac{j\eta d}{2\pi\rho_0 H_1^{(2)}(\underline{k}\rho_0)} \cdot H_0^{(2)}(\underline{k}\rho_{ij}), \tag{4.11}$$

where RW stands for *radial waveguide*, ρ_{ij} is the distance between ports, $\eta = \sqrt{\mu/\varepsilon}$, and $H_0^{(2)}$, $H_1^{(2)}$ are the Hankel functions of second kind and order 0 and 1, respectively. The Appendix A.4 includes the derivation of Eq. (4.11). For the calculation of self-impedances, the radius ρ_{ij} is evaluated at the port boundary ρ_0. Contrary to the cavity model, the port definition in the radial waveguide method is circular. The finite size of the second port for calculation of the transfer impedances is neglected in this formulation. A more rigorous formula that considers both port sizes has been proposed in the literature [110]. For the evaluated cases, it has been observed that Eq. (4.11) provides a very similar accuracy when compared to the improved formula.

To account for the effect of PMC boundaries on finite plates, the formulation can be extended by using image theory [102], as depicted in Figure 4.4. Here, the boundary condition is modeled using an infinite number of mirror sources to resemble the reflections at the cavity edges. The radial waveguide model combined with image theory (RW-IT) formula can be written as

$$Z_{ij}^{RW-IT}(\omega) = \frac{j\eta d}{2\pi\rho_0 H_1^{(2)}(\underline{k}\rho_0)} \sum_{r,s=-\infty}^{+\infty} H_0^{(2)}(\underline{k}\rho_{r,s}). \tag{4.12}$$

This formula requires the distances to all the image sources associated with one particular port, identified by the indexes r and s. The distances can be computed using the coordinate system reference shown in Figure 4.4 [102], with

$$\rho_{r,s} = \sqrt{x_r^{\,2} + y_s^{\,2}}, \tag{4.13}$$

and

$$x_r = x_i - r \cdot a - \begin{cases} x_j, & r \text{ even} \\ a - x_j, & r \text{ odd} \end{cases}, \tag{4.14}$$

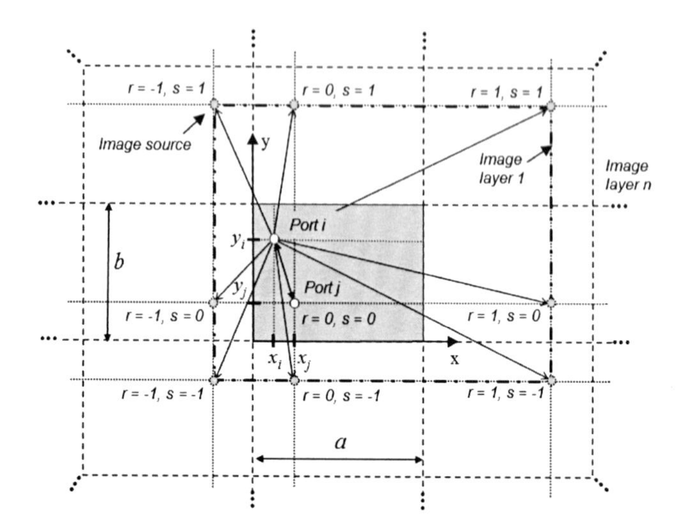

Figure 4.4 Radial waveguide approach combined with image theory to compute the parallel-plate impedance of finite rectangular plates and PMC boundaries. Diagram shows images associated with port *j* and distances to compute Z_{ij} for the first image layer.

$$y_s = y_i - s \cdot b - \begin{cases} y_j, & s \ even \\ b - y_j, & s \ odd \end{cases} . \quad (4.15)$$

The image configuration can be seen as a layered array, where the *n-th* layer is associated with the index sweeps $r,\ s\ =\ \pm\, n$, with one index fixed for $\pm n$ and the other sweeping, and vice versa, for a total of 4 passes. Each layer has $8{\cdot}n$ images. Figure 4.4 illustrates the first layer for port j, which is formed by 8 images.

It should be noted that both the cavity model and the radial waveguide method only consider propagating modes inside the cavity, assuming that the ports are electrically small and are located far away from each other. Multiple and back scattering between ports are not considered. Only two isotropic ports at a time are described without modeling their mutual interaction. If the coupling through the near-field region around ports or multiple scattering can not be neglected, e.g., for closely spaced ports, other formulations become necessary such as scattering methods [65].

The two discussed techniques involve infinite summations (Eqs. (4.5), (4.12)) that, in practice, must be truncated. The number of modes/images needed to estimate an accurate response and the convergence speed are issues of concern. These factors strongly depend on the geometrical characteristics of the structure, the frequency range of interest, and the losses, as will be shown in the next section.

4.3.3. Comparison of Convergence Properties and Computational Efficiency of the Methods

The two approaches discussed before have been applied in the literature to solve problems involving the modeling of power and ground planes for PCB and package structures, where good model-to-hardware correlation has been claimed. The formulae in Eq. (4.5) and Eq. (4.12) respond to different procedures and assumptions on the port definition, which make it difficult to derive a strict mathematical equivalence. However, the parallel-plate impedances computed by both techniques are consistent for several cases of practical interest. For instance, Figure 4.5 shows the impedance computed for a centered port on a rectangular lossy cavity, using Eqs. (4.5) and (4.12). Though the convergence behavior of each method is quite different, it can be observed that the results agree well. For this experiment, the rectangular port of the cavity resonator model corresponds to a rectangular region circumscribing the circular one.

In the cavity model, the series converges from lower towards higher frequencies; the first mode is related to the static plate capacitance and the following ones are associated with plate resonances. Although the contribution of higher order modes is more noticeable at higher frequencies, for lower frequencies they also affect the shape and location of the nulls of the response (series resonances) [105]. Many modes are required to achieve a good convergence and this number tends to increase with plane dimensions. Hence, the cavity model efficiency is negatively impacted for calculations involving large plates and high frequencies. For these cases, acceleration techniques [105]-[108] become indispensable to improve the convergence speed.

In contrast, the first iteration in the radial waveguide model provides a high frequency mean value (no reflections from edges). Then, as more images are included, the convergence starts improving towards lower frequencies. Images located further away reflect less power, due to the larger distance to the source and the losses. Since losses also increase with frequency, the contribution of a particular image will become negligible beyond a certain frequency. Thus, a few image layers could be sufficient to achieve good convergence if the distance to the image or the loss is large enough. Nonetheless, this technique converges very slowly when several image layers need to be considered, showing difficulties to predict accurately the low frequency part of the response.

The computation time of Z^{pp} as a function of the maximum mode index –for the cavity model– or number of images –for the radial waveguide method– is shown in Figure 4.6. Three different implementations of the cavity resonator model are compared: the double summation (CRM-DS, Eq. (4.5)), the single summation (CRM-SS, Eq. (4.10)), and the single summation improved for self-impedances (CRM-SSi) using the analytical series reported in [105].

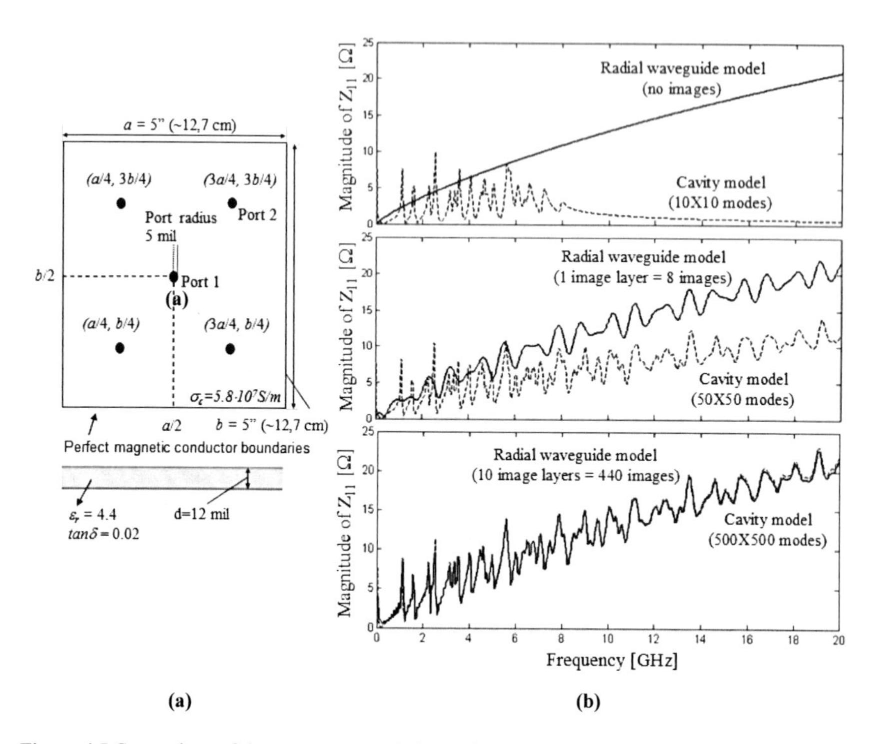

Figure 4.5 Comparison of the convergence of the methods to compute the parallel-plate impedance: (a) example of a single cavity with five circular ports, (b) port 1 self-impedance obtained by the cavity resonator model and the radial waveguide model for a different number of sum terms.

As depicted in Figure 4.6, the single-summation formulae can be computed relatively fast even when many modes are considered, with a comparable efficiency per iteration to the radial waveguide model (RW-IT, Eq. (4.12)). Ultimately, the overall efficiency will depend not only on the time per iteration, but mainly on the number of iterations required to achieve accurate results.

It is well known that the convergence for the cavity model depends strongly on port dimensions and their locations [105]. The smaller the ports and the separation among them, the slower the convergence will be. The behavior of the series was studied considering the structure in Figure 4.5(a), for a bandwidth from 0.1 up to 20 GHz and a convergence target of 0.001, measured as the maximum difference between two iterations of the algorithm (i.e. the sum truncated at *n-1* and *n*) for all the frequency points. For Z_{11}, the CRM-DS, CRM-SS and CRM-SSi series required a maximum mode index N equal to 801, 570, and 56, respectively. For the transfer impedances the

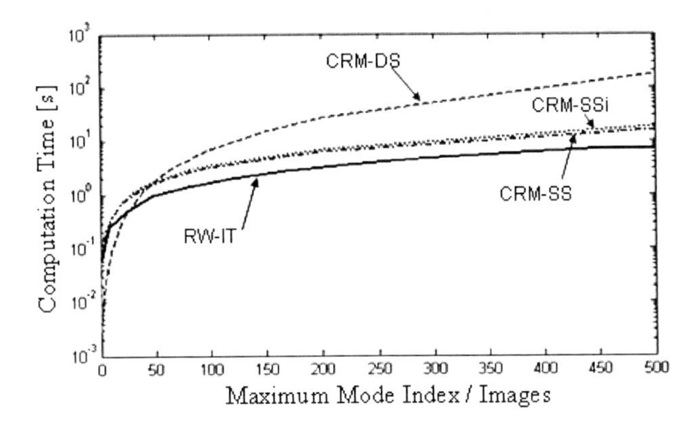

Figure 4.6 Computation time as a function of the number of iterations for the cavity resonator model: double summation (CRM-DS), single summation (CRM-SS), improved single summation (CRM-SSi), and the radial waveguide method (RW-IT). The computation time was obtained on a PC with 2.7 GHz, 2 GB RAM.

convergence can be much faster, but it depends on the relative port locations. For example, for Z_{12} (Figure 4.5(a)) and the same convergence criteria, the required N was equal to 60 for the CRM-DS, and equal to 39 for the single sums CRM-SS/SSi. The single summations help therefore to improve the efficiency by both reducing the time per iteration and the maximum number of iterations. Nevertheless, if either the maximum frequency of interest or the plate dimensions are increased, N and the computation time will tend to increase as well.

For the same bandwidth, the RW-IT requires thousands of images to achieve the convergence target. However, the required number of iterations decreases exponentially with frequency, and as the minimum frequency of interest increases, the method will become at some point faster than any cavity model implementation. For instance from 20 GHz on, less than 5 image layers (120 images) are required to compute Z_{11}. Although a few more image layers might be necessary depending on relative port positions, in comparison to the cavity model, it can be said that the convergence of this method is weakly influenced by port location or size.

As discussed in Section 3.3, shorted ports, which connect both plates of a cavity, can improve the electrical performance for signals transmitted over vias by offering additional return paths. After the calculation of Z^{pp} it is possible to model the effect of this type of port by assuming zero potential between the two plate locations being

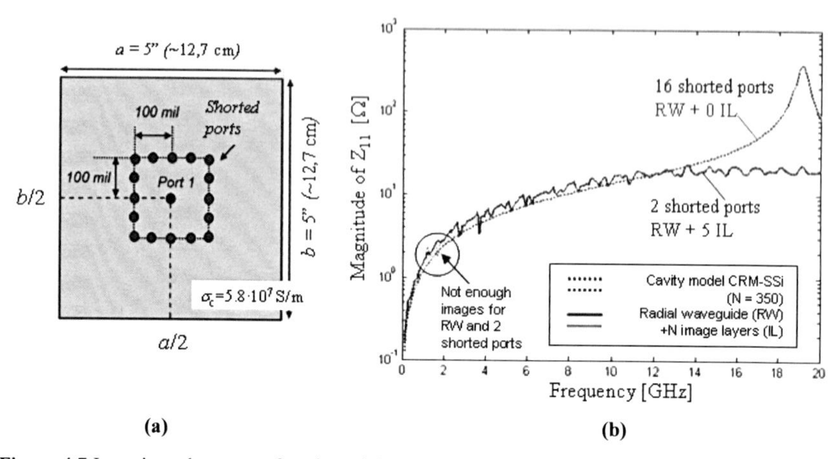

Figure 4.7 Input impedance as a function of the number of shorted ports. (a) Simulated structure and position of shorted ports (16 shown). (b) Computed results using the cavity model and the radial waveguide approach (curves almost overlapping). As the number of shorted ports increases less images are required in the computation of the radial waveguide method.

connected [16]. The radial waveguide, in contrast to the cavity resonator model, can take advantage of those ports since the summation can be truncated earlier without losing much accuracy. This idea is depicted in Figure 4.7, where a cavity was populated with a different number of shorted ports, starting from one up to sixteen. A smaller number of images becomes necessary to achieve good results with the radial waveguide method (within 5% deviation outside the resonances) as the number of shorted ports increases. After a certain number of shorted ports –sixteen in this example– it is not required to consider any image. On the other hand, for the computation with the cavity model enough modes are necessary in order to locate properly the resonances. The trend shown by the radial waveguide model is very important for the efficient computation of Z^{pp} in dense interconnect arrays, since it suggests that, if the number of shorted ports is high enough, the effect of either other ports placed further away or the board edges could be neglected.

4.3.4. Hybridization of Methods

From the discussion in Section 4.3.3, it can be stated that the cavity resonator method can be more efficient to compute small planes at relatively low frequencies, whereas the radial waveguide model shows important advantages for large lossy boards and high frequencies. These observations do not imply that for all cases one approach should be

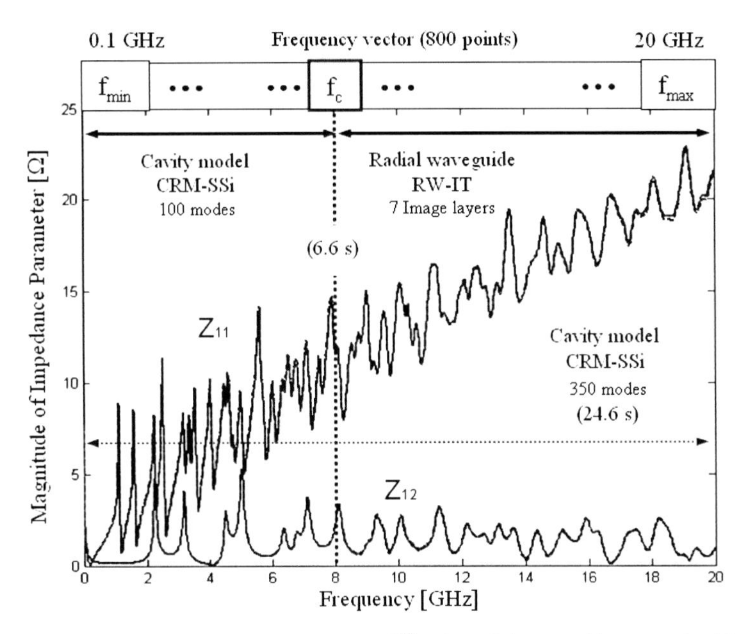

Figure 4.8 Hybridization of methods to compute Z^{pp}. The radial waveguide approach with image theory can compute the high frequency response part with better efficiency, whereas the cavity model is more effective to capture the resonant behavior at lower frequencies. Z_{11} and Z_{12} were computed, for the structure of Figure 4.5(a), using the hybrid method and compared to the cavity model single summation CRM-SSi (curves are almost completely overlapped).

selected with preference over the other one. The best of both techniques can be combined to efficiently compute Z^{pp} over a broad frequency range [18]. This is valid within the assumptions mentioned before regarding electrically small ports, sufficient separation and the neglect of multiple scattering. As illustrated in Figure 4.8 for the example in discussion (Figure 4.5(a)) and selecting a transition frequency of 8 GHz, the cavity model CRM-SSi was used to compute the low frequency part with $N = 100$. The high frequency part was computed by the radial waveguide technique with 7 image layers. The computation took 6.6 seconds for the five ports and 800 frequency points. In contrast, when only the CRM-SSi formula was applied for the entire calculation, $N = 350$ was required to achieve good agreement up to 20 GHz, for a total time of 24.6 seconds. In this case, the computation time was reduced by a factor of 3.7, however, if f_c and the number of modes are adjusted for every parameter independently, it is possible to further improve the efficiency. The relative computation time can also be reduced for larger plates, broader bandwidths, or higher losses, since the radial waveguide method will be able to compute the higher frequency range faster and the transition can be shifted towards lower frequencies.

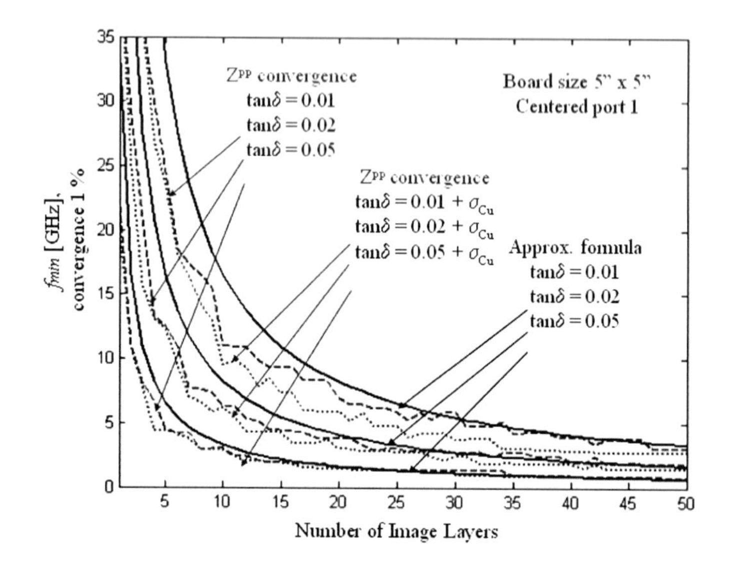

Figure 4.9 Minimum converged frequency *fmin* as a function of the number of image layers for the radial waveguide approach and different loss values, compared to the approximate formula in Eq. (4.17).

In order to find the transition frequency f_c, the mentioned convergence properties of the radial waveguide method can be used. The convergence can be related to the exponential decaying of the fields, which depends on the radial distance from the source and frequency dependent losses $\alpha(f)$. If the relative amplitude that reaches an image n, located at a distance l_n from the source, is lower than a desired error Δ (with $|\Delta|<1$), the following relation holds

$$\exp\left[-\alpha(f)\cdot l_n\right] \leq \Delta. \tag{4.16}$$

This means that any contribution coming from that image can be neglected, and the minimum frequency at which this relation is fulfilled becomes the smallest converged frequency f_{min}. Considering only dielectric loss $\alpha(f) \approx \pi f \sqrt{\mu_d \varepsilon_d} \cdot \tan\delta$, and a centered port on a square board, $l_n = n \cdot a$, the following approximate formula can be derived

$$f_c = f_{\min} \approx \frac{-\ln|\Delta|}{\pi \cdot \sqrt{\mu_d \varepsilon_d} \cdot n \cdot a \cdot \tan\delta}, \tag{4.17}$$

where n denotes the number of image layers. The formula can be numerically solved to include frequency dependent losses, and the distance a can be replaced by the minimum distance from the source point to an image layer n for non centered ports and transfer impedances. In these situations, to account for shorted ports, or for different convergence criteria, Eq. (4.17) can be empirically fitted. In Figure 4.9, the predicted f_{min} by Eq. (4.17) is compared to the one obtained from the direct calculation of Z^{pp} and a convergence target Δ of 1%. The results show that larger losses help to reduce f_{min} for a fixed number of images, since the fields decay more rapidly. The same trend was observed for an increasing distance to the cavity edges.

4.3.5. Limitations of Analytical Formulations and Outlook

The main advantage of the analytical formulations to compute Z^{pp} is that the numerical efficiency is usually very high in comparison to other techniques. The highest efficiency is achieved for the infinite plane case, since it does not require the evaluation of any series. It has been shown that this formulation can be useful for multiple via environments where the board edges play a marginal role.

In general, when the board edges cannot be neglected, the analytical formulations are limited to handle simple regular planes, mostly rectangles and some triangular and circular shapes [60],[111]-[113]. In order to handle irregular planes or plane perforations, such as split planes, it is possible to decompose the irregular board shape into regular pieces and to solve the total system in combination with segmentation and/or desegmentation techniques, e.g. in [114]-[116]. However, the evaluation made in [117] indicates that this is only efficient for relatively simple shapes and/or low frequencies, where the number of required interface ports and cavity modes are small. For more complex structures, the partitioning of the board becomes complex and the method efficiency may turn to be lower in comparison to other techniques like integral methods or full-wave simulations.

Another disadvantage of the analytical formulations is that the assumption that the ports are very small in comparison to the minimum wavelength of interest may not be fulfilled for increasing frequencies. It has been shown that approximations on the port shape can also be significant in the cavity model. To change the port definition would require reformulating the analytical expressions, which may be possible for relatively simple cases, but probably quite challenging for sources whose voltage and current distributions cannot be approximated as constant or averaged over the port region. As discussed before, the analytical formulations that have been treated cannot account for the near-field effects, and the multiple and back scattering among ports. Preliminary studies have shown that these factors may become important for very dense arrays and higher frequencies.

Numerical methods to compute Z^{pp} may be used to address these issues at the cost of a presumably lower numerical efficiency. The modeling of irregular shaped planes by the contour-integral method [17],[21] shows to be advantageous to overcome some of the limitations mentioned here, as it is going to be discussed in Chapter 6.

4.4. Computation of Via-to-Plane Capacitances

The available formulations for the parallel-plate impedance are usually not able to account for the near fields close to the via barrel and the clearance holes on planes (antipads). The work presented by Williamson in [97] shows that evanescent modes exist in this region and that they can be modeled as reactive elements. The via model approximates this near-field contribution by means of lumped capacitances, assuming that the antipads are still small in comparison to the wavelengths of interest. These via-to-plane capacitances can be computed either by means of quasi-static solvers, fitting formulas [94],[96], or closed-form expressions [118]. As depicted in Figure 4.10, the via-to-plane capacitance between the upper reference plane and the via segment is defined as

$$C_i^v = C_i^c + C_i^b + C_i^f . \tag{4.18}$$

where C_i^c is the coaxial capacitance between the plane arista and the via barrel, C_i^b the lateral capacitance between a plane side and the via barrel, and C_i^f the fringing capacitance between the upper/lower most plane and the via end.

The via-plane coaxial capacitance is calculated as [118]

$$C_i^c = \frac{2\pi\varepsilon_d t_c}{\ln(r_i^{ap} / r_i^v)} . \tag{4.19}$$

For adjacent cavities, the coaxial capacitance in Eq. (4.19) is considered only once. For thick reference planes it can be replaced by a coaxial transmission line segment, in order to approximate the introduced signal delay.

If the distance to plane edges or to other interconnects is sufficient, the via-plane lateral capacitance is computed from [118]

$$C_i^b = \frac{8\pi\varepsilon_d}{d \cdot \ln(r_i^{ap} / r_i^v)} \sum_{n=1,3,5\ldots}^{2N-1} \frac{1}{k_n^2 H_0^{(2)}(k_n r_i^v)} \cdot \left\{ \left[H_0^{(2)}(k_n r_i^{ap}) - H_0^{(2)}(k_n r_i^v) \right] \right\}, \tag{4.20}$$

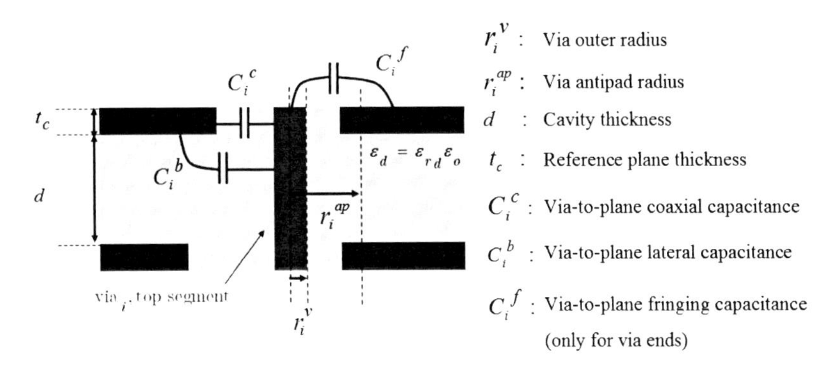

Figure 4.10 Definition of via-to-plane capacitances. The sketch considers just the capacitances between the top segment of a via and the upper reference plane.

with $k_n = \pm\sqrt{\omega^2 \mu_d \varepsilon_d - (n\pi / d)^2}$, r_i^v the via radius of the *i-th* via segment, r_i^{ap} the antipad radius, and $H_0^{(2)}$ the Hankel function of second kind and order zero. The series in Eq. (4.20) shows a fast convergence; for the study cases N = 20 was sufficient to achieve good results. The formula offers a reasonable estimate, usually within 10% in comparison to the extracted values by quasi-static commercial solvers. Typical via capacitance values per cavity sides range between 20 fF and 100 fF. The analytical formulae in [118] show that the capacitance values are frequency dependent, however for typical via geometries this dependency is weak below 40 GHz and it can be neglected without losing much accuracy. This property supports the premise that a quasi-static extraction serves as a reasonable approximation to calculate the via-to-plane capacitances.

Currently, there are no validated expressions to calculate via structures having arbitrary pad stacks, aligned to power planes and signal levels. For these cases, as well as for the aperture fields at via ends (modeled by C_i^f), the capacitances are computed by means of quasi-static or static solvers, e.g. [119]-[120].

The utilization of 3D or 2D methods able to extrapolate the results considering the cylindrical symmetry of the problem constitute attractive solutions that could be used to efficiently compute the capacitances of vias with arbitrary pad configurations. These alternatives require further research.

4.5. Modeling of Power and Ground Vias

One advantage of the via model topology is that it can be applied to handle signal and power/ground vias, in general called *PDN vias*. They serve to interconnect the reference planes assigned to the same potential or ground, to place PDN components such as decoupling capacitors, and to provide the power supply to other subsystems. These types of vias, as discussed in Section 3.3, can also improve the return path for signal vias.

The modeling of PDN vias requires the proper interpretation of the plane connectivity. When a PDN via is connected to a plane, the corresponding model entry associated with the via-to-plane capacitance must be replaced with a short circuit. For a single cavity, four alternatives are possible to describe a via segment depending on the plane assignment, as depicted in Figure 4.11. If the PDN via does not contact any of the planes, the via is handled as a signal via and the two capacitance entries are defined (case 1). If only one plane is contacted, the respective entry is replaced by a short circuit (cases 2, 3). If the via and both planes of the cavity are assigned to the same PDN net, both capacitances are substituted by short circuits (case 4). Since the models neglect the field penetration in the reference planes and the inter-level crosstalk at the board edges, the cavities are only coupled to each other through the vias.

If the via is touching both planes (case 4 in Figure 4.11), it can be reduced right after the 2D computation of Z^{pp} –i.e. before expanding the matrix according to Eq. (4.3). The assumption made is that the voltage across the via is equal to zero, thus the impedance matrix can be rearranged by grouping the shorted ports as follows [16]

$$\begin{bmatrix} \overline{V}_s \\ 0 \end{bmatrix} = \underbrace{\begin{bmatrix} \overline{\overline{Z}}{}^{pp}_{ss} & \overline{\overline{Z}}{}^{pp}_{sg} \\ \overline{\overline{Z}}{}^{pp}_{gs} & \overline{\overline{Z}}{}^{pp}_{gg} \end{bmatrix}}_{\overline{\overline{Z}}{}^{pp}_{t}} \cdot \begin{bmatrix} \overline{I}_s \\ \overline{I}_g \end{bmatrix}. \tag{4.21}$$

The subscripts s and g refer to the signal (non-shorted) and ground via (shorted) vectors, respectively. The reduced impedance matrix then contains only the signal vias from Eq. (4.21)

$$\overline{\overline{Z}}{}^{pp}_{t'} = \overline{\overline{Z}}{}^{pp}_{ss} - \overline{\overline{Z}}{}^{pp}_{sg} \cdot \left(\overline{\overline{Z}}{}^{pp}_{gg} \right)^{-1} \cdot \overline{\overline{Z}}{}^{pp}_{gs}, \tag{4.22}$$

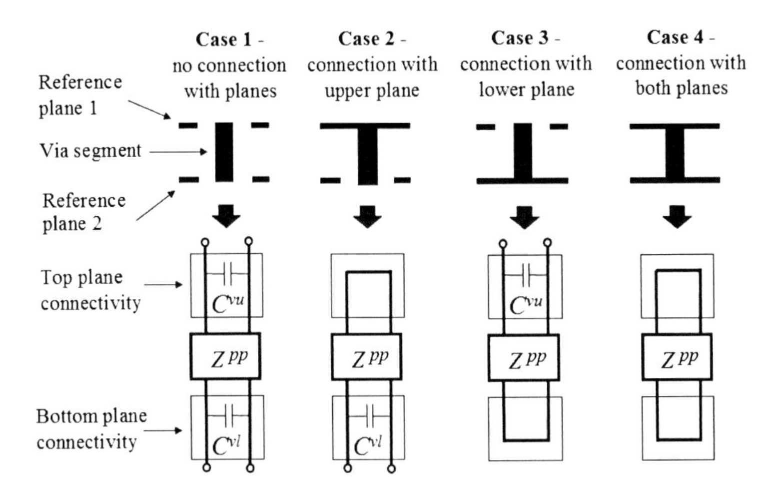

Figure 4.11 Via connectivity with respect to the reference planes for a single cavity. For via transitions crossing the plane through a clearance hole (antipad) the model entry corresponds to the via-to-plane capacitance. For vias that are connected to the reference planes, the respective model entry is a short circuit.

which is the Schur complement of $\overline{\overline{Z}}_t^{pp}$. The size of the reduced matrix is given by the number of non-shorted ports denoted with the subscript s.

The aforementioned transformation can also be applied to the expanded matrix, with the consideration that it can be only transformed to Z-parameters after the capacitances have been included, since Eq. (4.3) becomes singular for Z-parameters. Alternatively, the reduction of ground vias can also be accomplished by terminating the corresponding port with a short circuit in any network parameter form.

4.6. Extended Model for Vias and Traces

Signal vias in high-speed electronic systems are usually connected to traces of internal layers. Therefore, the extension of the via model to handle the via-stripline transition was necessary and it has been developed as part of this work [15],[19] (Figure 4.12). The via-trace transition is not trivial for reference planes that support parallel-plate modes. For this scenario, the signals traveling on vias can be coupled into both the modes guided between the parallel plates and the transmission line modes associated with the traces (Figure 4.13).

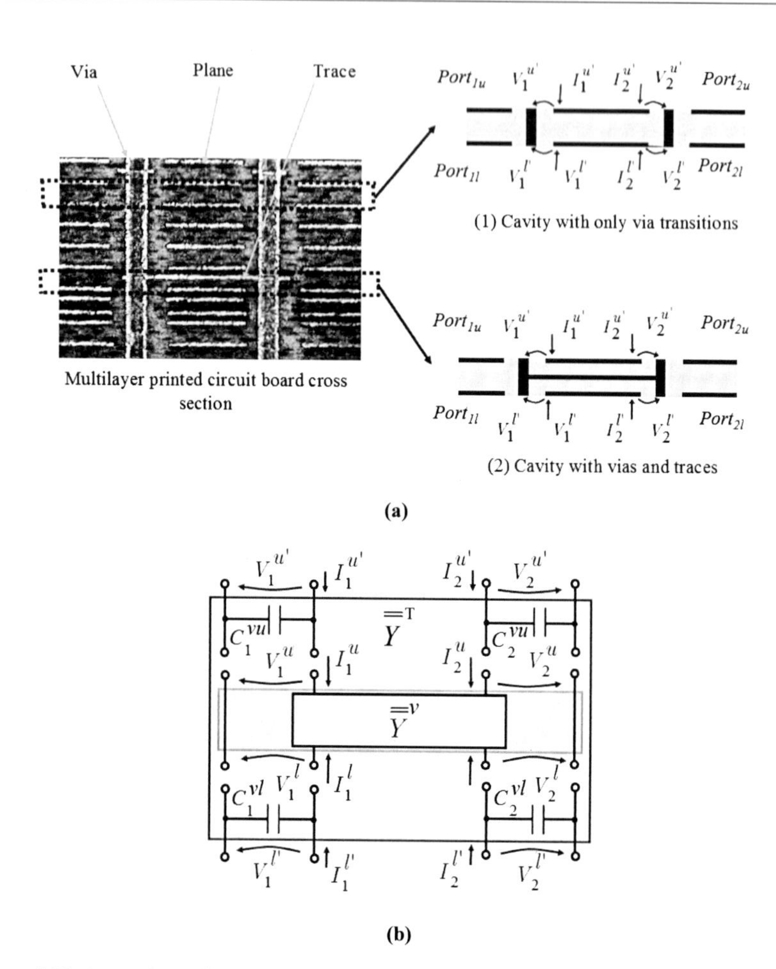

Figure 4.12 Approach used to model multilayer board structures with vias and internal traces connecting vias. (a) Each cavity is modeled separately for both layers with only via transitions (1), or with vias and stripline transitions (2). (b) General equivalent network representation of the model, where the matrix $\overline{\overline{Y}}^{v}$ corresponds to the parallel-plate model when only via transitions are present, or to the superposition of the parallel-plate and transmission line models when vias and traces are present.

Assuming that the parallel-plate modes and the transmission line modes are only coupled at the via locations, it is possible to compute each set of modes separately and apply modal decomposition [121]-[122] to combine them by selecting suitable transformation matrices.

4.6.1. Modal Decomposition

The description of a stripline routed between non-ideal reference planes can be found in the literature [79]. The process involves the transformation of the terminal transmission line equations into a set of modal uncoupled equations in the frequency domain. An n-conductor system can be described by the following general expressions

$$\frac{\partial}{\partial z}\overline{V}(z) = -\overline{\overline{Z}}\cdot\overline{I}(z), \tag{4.23}$$

$$\frac{\partial}{\partial z}\overline{I}(z) = -\overline{\overline{Y}}\cdot\overline{V}(z), \tag{4.24}$$

with $\overline{\overline{Z}} = \overline{\overline{R}} + j\omega\overline{\overline{L}}$, $\overline{\overline{Y}} = \overline{\overline{G}} + j\omega\overline{\overline{C}}$. The terminal voltages and currents can be written in terms of modal voltages and currents by defining the transformation matrices $\overline{\overline{T_v}}$ and $\overline{\overline{T_i}}$

$$\overline{V}(z) = \overline{\overline{T_v}}\cdot\overline{V}_m(z), \tag{4.25}$$

$$\overline{I}(z) = \overline{\overline{T_i}}\cdot\overline{I}_m(z). \tag{4.26}$$

By substituting Eqs. (4.25) and (4.26) into (4.23) and (4.24), the transmission line equations for the modal quantities (m) become

$$\frac{\partial}{\partial z}\overline{V}_m(z) = -\overline{\overline{T_v}}^{-1}\overline{\overline{Z}}\,\overline{\overline{T_i}}\cdot\overline{I}_m(z), \tag{4.27}$$

$$\frac{\partial}{\partial z}\overline{I}_m(z) = -\overline{\overline{T_i}}^{-1}\overline{\overline{Y}}\,\overline{\overline{T_v}}\cdot\overline{V}_m(z). \tag{4.28}$$

Assuming a homogeneous dielectric material and perfect conducting planes, Eqs. (4.27) and (4.28) are diagonalized if the inductance matrix $\overline{\overline{L}}$ is diagonalized ([45],[79], see Appendix A.5)

$$\overline{\overline{T_v}}^{-1}\overline{\overline{L}}\,\overline{\overline{T_i}} = \overline{\overline{L_m}} = \begin{pmatrix} \overline{\overline{L_{pp}}} & 0 \\ 0 & \overline{\overline{L_{tl}}} \end{pmatrix}, \tag{4.29}$$

where $\overline{\overline{L_{pp}}}$ stands for the inductance of the parallel planes and $\overline{\overline{L_{tl}}}$ for the inductance of the trace.

It can be shown (Appendix A.5) that the following transformation matrices can be used to couple or decouple the two modes at every via location [79]

$$\begin{pmatrix} \phi_i^u - \phi_i^l \\ \phi_i^s - \phi_i^l \end{pmatrix} = \begin{pmatrix} 1 & 0 \\ -k_i & 1 \end{pmatrix} \begin{pmatrix} V_i^{pp} \\ V_i^{tl} \end{pmatrix}, \tag{4.30}$$

$$\begin{pmatrix} I_i^u \\ I_i^s \end{pmatrix} = \begin{pmatrix} 1 & k_i \\ 0 & 1 \end{pmatrix} \begin{pmatrix} I_i^{pp} \\ I_i^{tl} \end{pmatrix}. \tag{4.31}$$

The superscripts u and l denote the *upper* and *lower* planes of a cavity; the superscripts pp and tl are related to the *parallel-plate* mode, and the trace *transmission line* mode, respectively. The definition of voltages and currents is included in Figure 4.13. Neglecting the trace thickness, a simple formula for k_i can be obtained in terms of a dielectric height ratio [79] (see Appendix A.5)

$$k_i = -\frac{h_i^l}{h_i^l + h_i^u}. \tag{4.32}$$

The factor k_i should not be confused with the wave number $\underline{k}$. In case of a non-zero trace thickness, this factor can also be numerically computed. It has been demonstrated that Eq. (4.32) offers a good approximation for most practical cases of interest [123]. For typical offset striplines the value of k_i obtained with Eq. (4.32) and the numerical extraction differs just slightly (below 5%), whereas for centered and symmetric cases both solutions compute almost the same value.

The matrices in Eqs. (4.30) and (4.31) can be explicitly written in order to formulate an equivalent circuit

$$I_i^u = I_i^{pp} + k_i I_i^{tl}, \qquad I_i^s = I_i^{tl}, \tag{4.33}$$

$$\phi_i^u - \phi_i^l = V_i^{pp}, \qquad \phi_i^s - \phi_i^l = -k_i \cdot V_i^{pp} + V_i^{tl}. \tag{4.34}$$

Equation (4.33) states that the current through the upper plane is equal to the sum of the currents from the parallel-plane mode, and those from the transmission line mode multiplied by the factor k_i. This factor gives the portion of the trace return current that flows through the upper and lower planes. Equation (4.33) can be represented by a current-controlled current-source as shown in Figure 4.13(b). Similarly, the voltage between the trace and the lower plane is also dependent on k_i. It is equal to the voltage drop due to the trace mode minus the drop between the planes owed to the parallel-plate mode, multiplied again by the factor k_i. Equation (4.34) can be described by a

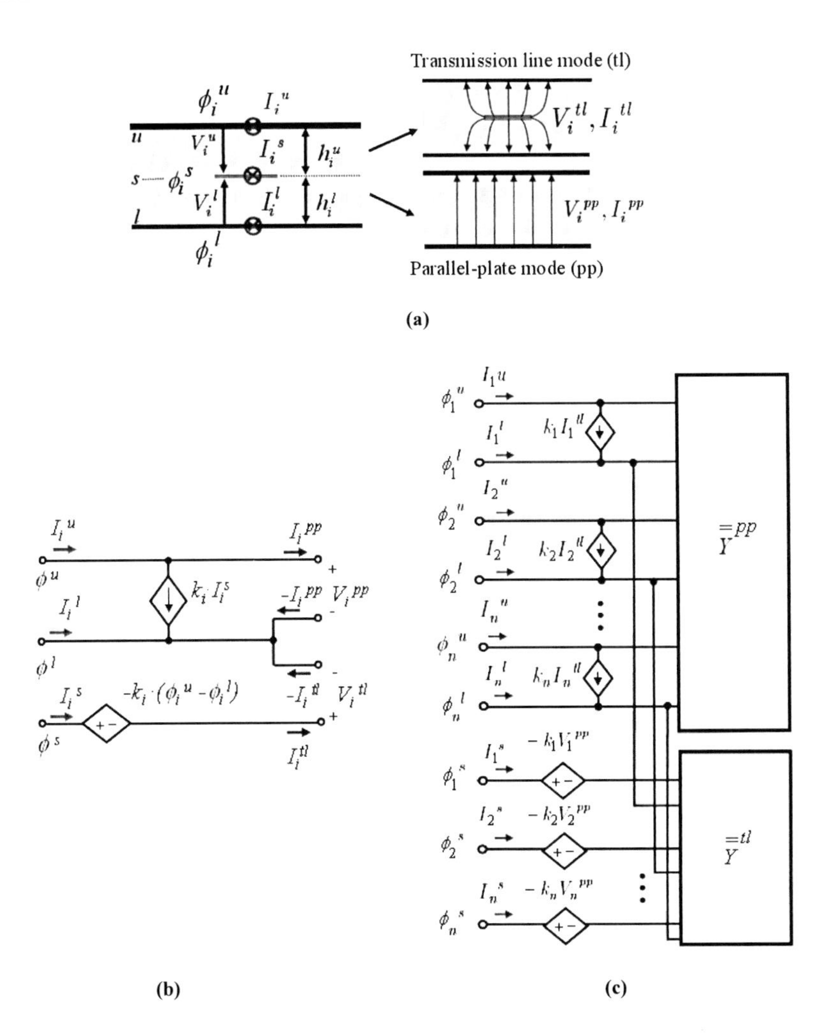

Figure 4.13 Illustration of the modal decomposition approach applied to model vias connected by traces. (a) Terminal and modal parameter definition. (b) Equivalent circuit for a single via-trace transition. (c) Equivalent circuit extended to n-ports. The blocks $\overline{\overline{Y}}^{pp}$ and $\overline{\overline{Y}}^{tl}$ correspond to the parallel-plane and the trace transmission line models as Y-parameters.

voltage-controlled voltage-source. The equivalent circuit can be extended for n-ports as shown in Figure 4.13(c). The basic transformation cell of Figure 4.13(b) is used to couple the different ports of the parallel-plate and transmission line models to the corresponding via-to-stripline transition terminal quantities.

4.6.2. Via Model Including Stripline Transitions

The via model in Section 4.2 has been extended in this work to include traces connecting vias [15],[19]. The network topology remains unchanged, but the contribution of the transmission line model for traces is incorporated by applying the modal decomposition technique discussed in the previous section. The formulation as a stand-alone network block, as shown in Figure 4.12(b), requires writing the model in terms of voltages and currents defined at each cavity side. The expressions in Eq. (4.30) and (4.31) can be manipulated to solve a system of equations for the upper and lower voltages with respect to the signal (via) terminal $V_i^{\mathrm{u}} = \phi_i^{\mathrm{u}} - \phi_i^{\mathrm{s}}$, $V_i^{l} = \phi_i^{l} - \phi_i^{\mathrm{s}}$, and the currents I_i^{u}, I_i^{l}. The following Y-parameter expression is obtained ([19], see Appendix A.6)

$$\begin{pmatrix} \overline{I}^{u} \\ \overline{I}^{l} \end{pmatrix} = \underbrace{\begin{pmatrix} k^2 \cdot \overline{\overline{Y}}^{tl} + \overline{\overline{Y}}^{pp} & (-k^2 - k) \cdot \overline{\overline{Y}}^{tl} - \overline{\overline{Y}}^{pp} \\ (-k^2 - k) \cdot \overline{\overline{Y}}^{tl} - \overline{\overline{Y}}^{pp} & (k^2 + 2k + 1) \cdot \overline{\overline{Y}}^{tl} + \overline{\overline{Y}}^{pp} \end{pmatrix}}_{\overline{\overline{Y}}^{V}} \cdot \begin{pmatrix} \overline{V}^{u} \\ \overline{V}^{l} \end{pmatrix}. \tag{4.35}$$

The matrices for the parallel-plate and trace (stripline) model are defined respectively as

$$\overline{I}^{pp} = \overline{\overline{Y}}^{pp} \cdot \overline{V}^{pp}, \tag{4.36}$$

$$\overline{I}^{tl} = \overline{\overline{Y}}^{tl} \cdot \overline{V}^{tl}. \tag{4.37}$$

The $\overline{\overline{Y}}^{tl} = \overline{\overline{Z}}^{tl^{-1}}$ matrix contains the transmission line admittance matrix assuming ideally grounded planes. Taking the admittance matrix in Eq. (4.35) and adding the via-to-plane capacitance term in Eq. (4.4), a generalized model can be formulated for one cavity with n vias and m traces. As Y-parameters the block matrices can be written as

$$\begin{aligned} \overline{\overline{Y}}^{T} = & \begin{pmatrix} \overline{\overline{Y}}^{cu} & \overline{\overline{0}} \\ \overline{\overline{0}} & \overline{\overline{Y}}^{cl} \end{pmatrix} + \begin{pmatrix} +\overline{\overline{Y}}^{pp} & -\overline{\overline{Y}}^{pp} \\ -\overline{\overline{Y}}^{pp} & +\overline{\overline{Y}}^{pp} \end{pmatrix} \\ & + \begin{pmatrix} -k \cdot \overline{\overline{Y}}^{tl} & 0 \\ 0 & (k+1) \cdot \overline{\overline{Y}}^{tl} \end{pmatrix} + \begin{pmatrix} (k^2 + k) \cdot \overline{\overline{Y}}^{tl} & -(k^2 + k) \cdot \overline{\overline{Y}}^{tl} \\ -(k^2 + k) \cdot \overline{\overline{Y}}^{tl} & (k^2 + k) \cdot \overline{\overline{Y}}^{tl} \end{pmatrix}. \end{aligned} \tag{4.38}$$

Equation (4.38) describes the general network-level building block, depicted in Figure 4.12(b) for four ports. The contribution of four different components can be identified, namely, the via-to-plane capacitances, the parallel-plate impedance, the trace transmission line model –expressed as two lines coupled with the top and bottom planes, respectively, and related by the factor k– and a fourth factor which results from the modal transformation on the parallel-plate impedance and the transmission line mode.

For vias not connected by traces, the entries related to the transmission line model (third and fourth terms in Eq. (4.38)) vanish and the expression is reduced to Eq. (4.4) for such elements. For simplicity, the factor k is assumed to be equal for all the elements inside the cavity. However, if this is not the case, the factor can be calculated independently for the different ports of the structure and k becomes a matrix. This makes it possible to solve cases with more than one signal layer routed between two reference planes, provided that the coupling between the layers is properly mapped in the transmission line model and the mode conversion between the parallel-plate and trace modes mostly occurs at via locations.

In parallel to the development of this work, a similar formulation to the one used here was presented by X. C. Wei *et al.* in [73]. However, in their solution the fourth term in Eq. (4.38) is neglected. It has been concluded that the complete expression is more accurate at higher frequencies. For instance, for the case discussed later in Section 5.2.2, the effect of the fourth term starts to be noticeable for frequencies above 6 GHz. When this term is neglected, the magnitude of the S-parameters can show a deviation in the range of 2 dB and more.

For a simple case of two via transitions crossing a single cavity and connected by a centered trace (k = -0.5) as shown in Figure 4.14, Eq. (4.38) becomes

$$\begin{pmatrix} I_1^{u'} \\ I_2^{u'} \\ I_1^{l'} \\ I_2^{l'} \end{pmatrix} = \underbrace{\begin{pmatrix} \overline{\overline{Y}}^{cu} & \overline{\overline{0}} \\ \overline{\overline{0}} & \overline{\overline{Y}}^{cl} \end{pmatrix} + \begin{pmatrix} (0.25)\overline{\overline{Y}}^{tl} + \overline{\overline{Y}}^{pp} & (0.25)\overline{\overline{Y}}^{tl} - \overline{\overline{Y}}^{pp} \\ (0.25)\overline{\overline{Y}}^{tl} - \overline{\overline{Y}}^{pp} & (0.25)\overline{\overline{Y}}^{tl} + \overline{\overline{Y}}^{pp} \end{pmatrix}}_{\overline{\overline{Y}}^{T}} \cdot \begin{pmatrix} V_1^{u'} \\ V_2^{u'} \\ V_1^{l'} \\ V_2^{l'} \end{pmatrix}. \tag{4.39}$$

Both the parallel-impedance and stripline matrices are 2-port networks with their terminals defined at the via locations. For the plane model the matrix can be computed using, for instance, the formulae in Section 4.3. The resultant 2D matrix (2x2) has the form

$$\begin{pmatrix} I_1^{pp} \\ I_2^{pp} \end{pmatrix} = \begin{pmatrix} Y_{11}^{pp} & Y_{12}^{pp} \\ Y_{21}^{pp} & Y_{22}^{pp} \end{pmatrix} \cdot \begin{pmatrix} V_1^{pp} \\ V_2^{pp} \end{pmatrix}, \tag{4.40}$$

with $\overline{\overline{Y}}^{pp} = \overline{\overline{Z}}^{pp^{-1}}$. For the stripline, the 2D matrix is defined as

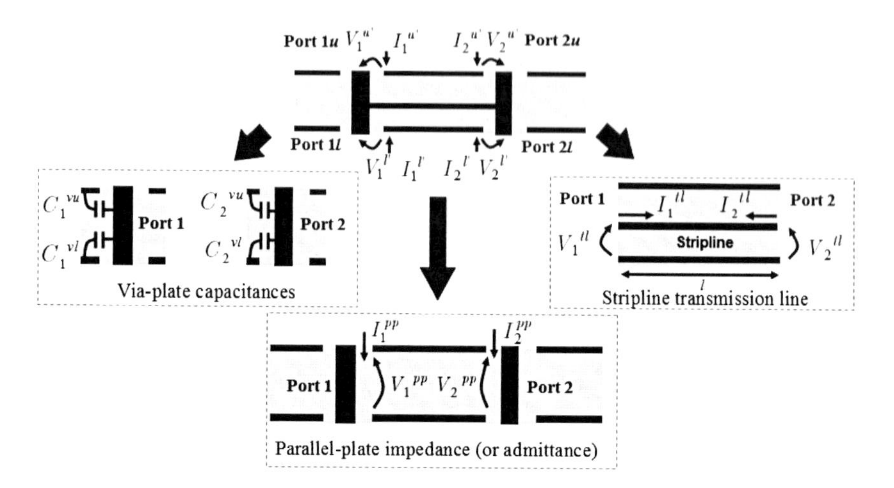

Figure 4.14 Example of two vias connected by a centered stripline for generation of the model component matrices.

$$\begin{pmatrix} I_1^{tl} \\ I_2^{tl} \end{pmatrix} = \begin{pmatrix} Y_{11}^{tl} & Y_{12}^{tl} \\ Y_{21}^{tl} & Y_{22}^{tl} \end{pmatrix} \cdot \begin{pmatrix} V_1^{tl} \\ V_2^{tl} \end{pmatrix}, \tag{4.41}$$

or [124]

$$\begin{pmatrix} I_1^{tl} \\ I_2^{tl} \end{pmatrix} = \frac{1}{Z_o} \cdot \begin{pmatrix} \coth(\gamma l) & -1/\sinh(\gamma l) \\ -1/\sinh(\gamma l) & \coth(\gamma l) \end{pmatrix} \cdot \begin{pmatrix} V_1^{tl} \\ V_2^{tl} \end{pmatrix}, \tag{4.42}$$

with l the length of the stripline, Z_o the characteristic impedance of the line and γ the propagation constant. Z_o and γ can be computed analytically for a homogeneous cross section [53]. The numerical computation of the "per-unit-length" parameters is another alternative to solve the problem [122]-[123].

For more elements the computation of Z^{pp} is identical; however the computation of the Y^{tl} matrix for multiple coupled traces (e.g. differential striplines) requires the solution of a multiconductor transmission line system (MTL). These models can be computed with any program able to solve MTL configurations such as 3D numerical solvers, 2D solvers if a uniform cross section serves as a good approximation [10],[19], or closed-form expressions for simple configurations when available.

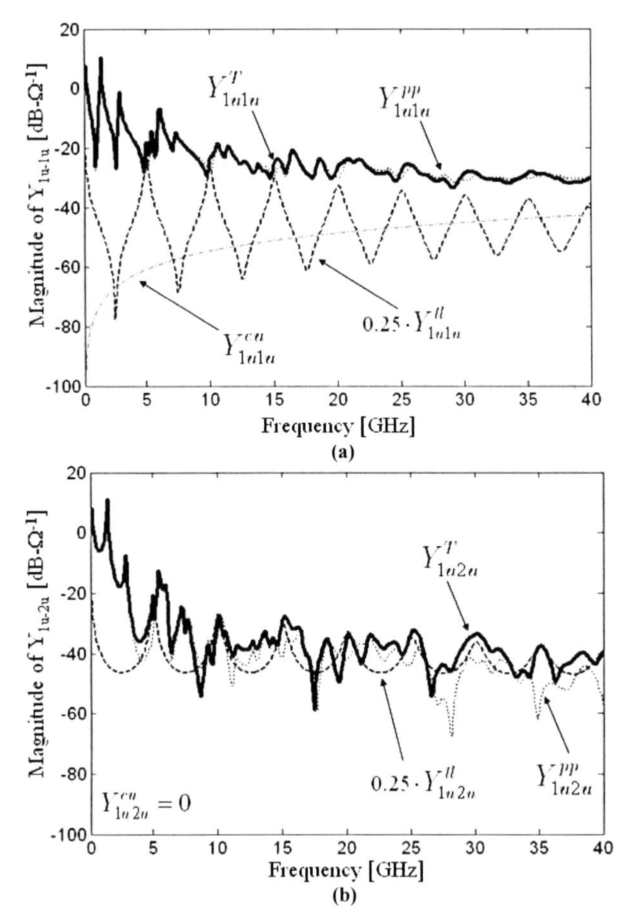

Figure 4.15 Contribution of model components for the case shown in Figure 4.14. (a) Self-admittance of the first upper port. (b) Transfer admittance between the two upper ports. The following parameters are used: via radius 5 mil, antipad radius 15 mil, cavity thickness 12 mil, ε_r = 3.8, tanδ = 0.03, plane size 1.2 x 1.2 inch, port 1 at (0.6,0.2) inch, port 2 at (0.6,1) inch, PMC boundaries, trace length 600 mil, trace width 4 mil, trace thickness 1 mil, k = -0.5.

Finally, the capacitances are computed or extracted, and arranged in the following form

$$\overline{\overline{Y}}^{cu} = \begin{pmatrix} j\varpi C_1^{vu} & 0 \\ 0 & j\varpi C_2^{vu} \end{pmatrix}, \tag{4.43}$$

$$\overline{\overline{Y}}^{cl} = \begin{pmatrix} j\varpi C_1^{vl} & 0 \\ 0 & j\varpi C_2^{vl} \end{pmatrix}. \tag{4.44}$$

Figure 4.15 depicts the contribution of each model component for the self-admittance of the first upper port and the transfer admittance between the two upper ports defined in Figure 4.14. The considered structure has only one cavity and therefore the contribution of the via capacitance is not very significant. The effect of the trace is more pronounced for the transfer admittance.

4.7. Generalized Method for Simulation of Multilayer Substrates

The presented models for vias and traces can be applied to compute the response of each cavity in a multilayer substrate with n-vias and m-traces. Next, the partial results obtained for the cavities are merged to get the final response at the defined via ports, for instance as a S-parameter matrix. Note that the models are intended to describe only the region of the substrate enclosed by solid reference planes.

ABCD (or chain) matrices can be efficiently concatenated by matrix multiplication. However, this procedure is only applicable to cases with signal vias or with power vias shorted at both cavity sides. The *ABCD* form fails in more complicated cases having buried/ blind vias and mixed reference planes [16], since the chain parameters for some ports can not be defined.

The segmentation method [78] offers a general solution to combine network blocks with an arbitrary definition and number of ports, as depicted in Figure 4.16. The prerequisite to apply this technique is the proper ordering of ports. The results computed for adjacent layers can be arranged in terms of connected and non-connected ports. Then, the matrices are merged by ensuring voltage and current continuity at the connected ports; the output is an equivalent reduced matrix with only the non-connected ports. Expressions to perform this operation are available in terms of S-, Z-, or Y-parameters [78] (Appendix A.7). Although the numerical efficiency of the segmentation process is higher in terms of impedance parameters [125], for complex structures which may include PDN vias shorting the planes, open terminations or loads, the segmentation in terms of S-parameters shows better numerical stability [16],[19].

In order to map power/ground vias into the models, the matrices containing the via-to-plane capacitances need to be extended to consider the connectivity of the vias as well, as mentioned in Section 4.5. For this purpose, the first matrix term in Eq. (4.38) is handled separately to form interconnection layers, as shown in Figure 4.17. In case the PDN via is connected to the plane, the via-to-plane capacitance entry is replaced with a short circuit. For via ends not defined as ports, the corresponding entry can be either terminated with the via-to-plane capacitance or with an open circuit. The

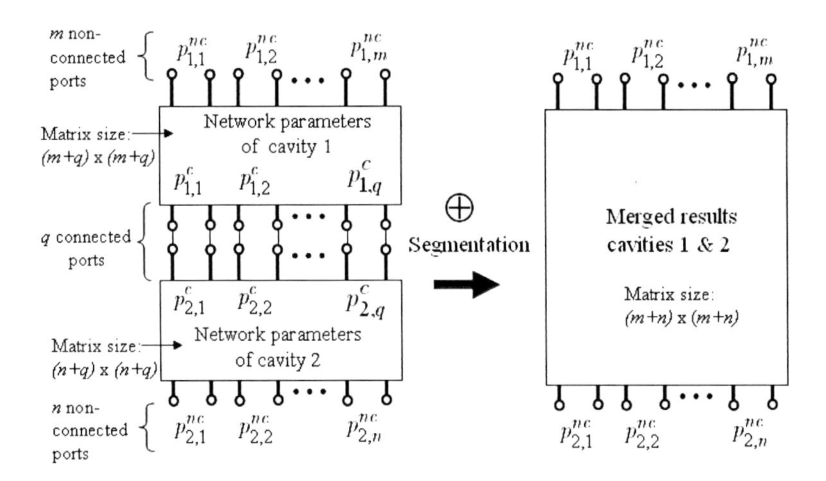

Figure 4.16 Concatenation of partial cavity results by using segmentation. The ports of the network description for different cavities are arranged in connected *(c)* and non-connected ports *(nc)*. The segmentation method reduces the connected ports and returns an equivalent network with only the non-connected ports. Appendix A.7 provides the formulae for segmentation.

transmission line model for the stripline and the plane model are included in the matrices Y^v, according to Eq. (4.35).

The partitioning approach and the blocks to be merged by using segmentation techniques are illustrated in Figure 4.17 with a simple two-layer example containing signal, ground and power vias. Two cavities and four interconnection layers are generated as partial results. The segmentation procedure is applied on these blocks in order to merge them and to obtain the final response. The total result should have the form of a 2-times.2 matrix per frequency, defined for the two ports at the top side (ports *1u* and *2u*).

Figure 4.18 shows a simplified functional block diagram of the simulation framework based on the proposed models. As part of this project, a code for automated simulation of multilayer substrate has been developed (Appendix B), which was used to generate all the model results included in the next chapters. In contrast to previous work on via modeling [94]-[96], the simulation process does not rely on the utilization of SPICE-like network simulators. As a consequence, the circuit topology at "wire" level is not required for each configuration to be simulated. The presented approach is based on microwave network parameters, which allows the automatic generation of the models.

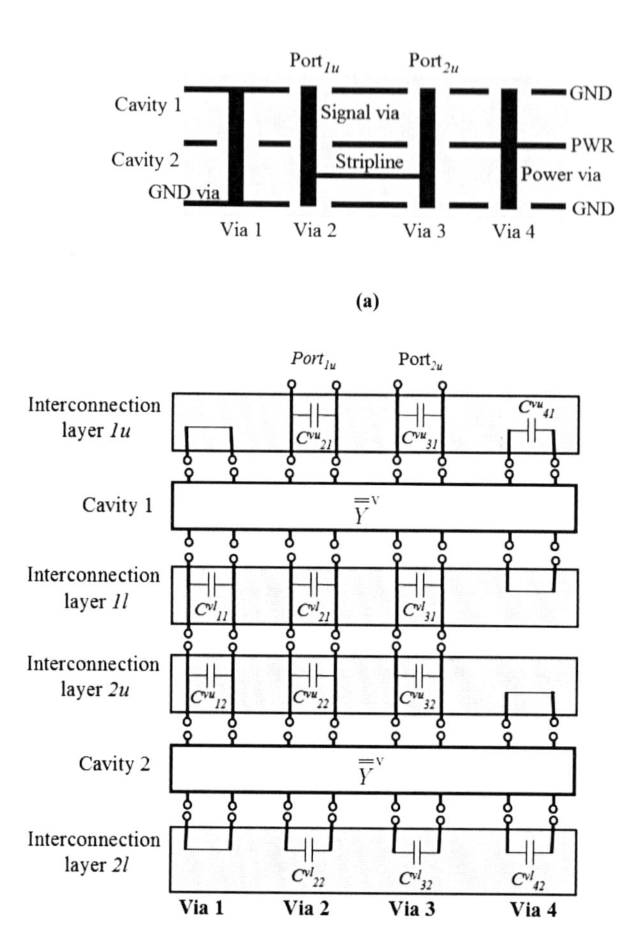

Figure 4.17 Partitioning approach to model multilayer structures. (a) Two-cavity example with mixed reference planes and power/ground vias. (b) Segmentation approach exemplified with this two-cavities four-via case.

This is crucial for rapid analysis of complex arrays and for opening the possibility of applying optimization techniques. The developed code makes use of high-level text input files that contain the description of the structure to be simulated. This information is read by the program to automatically generate the data structures and to calculate the overall response.

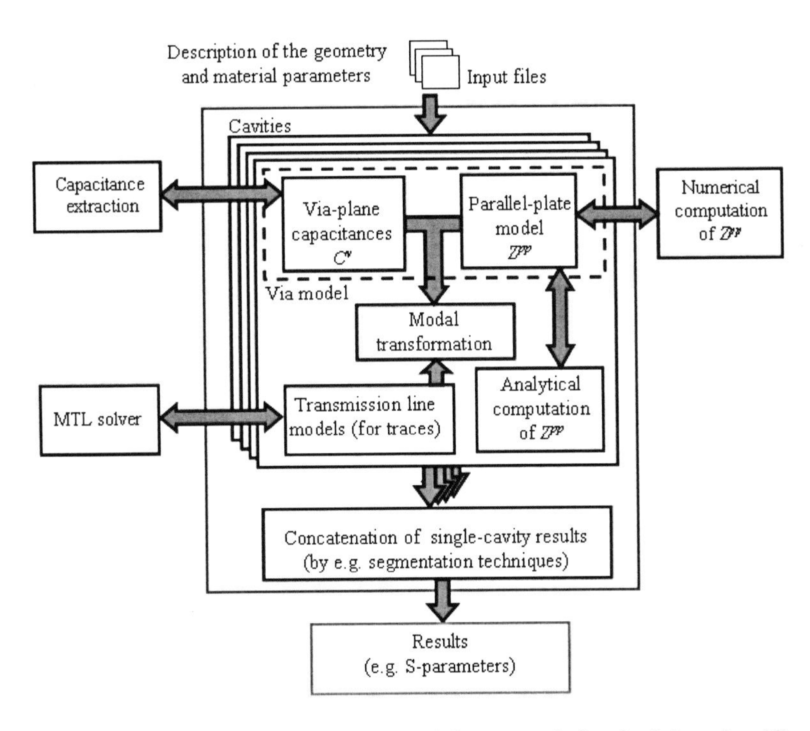

Figure 4.18 Functional block diagram of the modeling approach for simulation of multilayer substrates. The figure depicts the interrelation between the different model components.

In summary, the general method depicted in Figure 4.18 comprises the computation of the model for vias and traces of each single cavity. The parallel-plate impedance can be calculated analytically or externally by numerical methods in case of, for instance, irregular plates. The via-to-plane capacitances are obtained analytically if no pads are defined. Extracted values are imported if a pad-stack definition exists. Terminations and lumped elements are incorporated into the same interconnection matrices with the via-to-plane capacitances. For traces, simple analytical models can be internally generated; an external solver is required for arbitrary MTL systems. The code is able to import either 3D or 2D models. An automatic interface with the 2D-MTL solver in CONCEPT-II [126] is available [10],[123]. The modal decomposition method is used to combine the plane and trace models. Once the cavities have been computed these partial results are concatenated. This task can be done sequentially or in parallel depending on the available computing resources and the complexity of the structure.

5. Validation of the Models with Full-Wave Simulations and Measurements

In this chapter the utilization of the proposed models is discussed, oriented to the simulation of interconnect structures typically found in multilayer substrates such as single-ended and differential links with diverse power/ground via configurations. The code developed in this work (Appendix B) has been applied to simulate many configurations and the results have been validated against full-wave methods and measurements. The scope of the simulations and their numerical efficiency are discussed in combination with the selected examples. Model efficiency, limitations, and perspectives for further developments are provided in the last sections.

The test vehicles and measured data presented in this chapter were provided by the High-Speed I/O Subsystems and Packaging Group at IBM T. J. Watson Research Center, Yorktown Heights, New York, USA.

5.1. Multilayer Vias

The first correlation studies [16] were done with structures having only vias, in order to validate the existing via formulation in their microwave network parameter form. The results were compared with the ones reported in previous publications, which relied on the utilization of SPICE-like network simulators to implement the via model [71],[80]. Measurements and full-wave analyses were also used for the validation.

In Figure 5.1(a), a case study with two through-hole signal vias crossing six ground planes is presented. It is a configuration comparable to the one previously analyzed in [94] (Figure 5.1(c)). PEC boundaries were defined in order to have a clearly delimited electromagnetic environment. This condition was realized on test vehicles by using ground via cages [94]. The obtained results are compared in Figure 5.2. There is good agreement between the results of the full-wave FEM simulations [127] and the models expressed as microwave network parameters up to 20 GHz. The response can also be predicted with less accuracy at higher frequencies. The results from measurement and

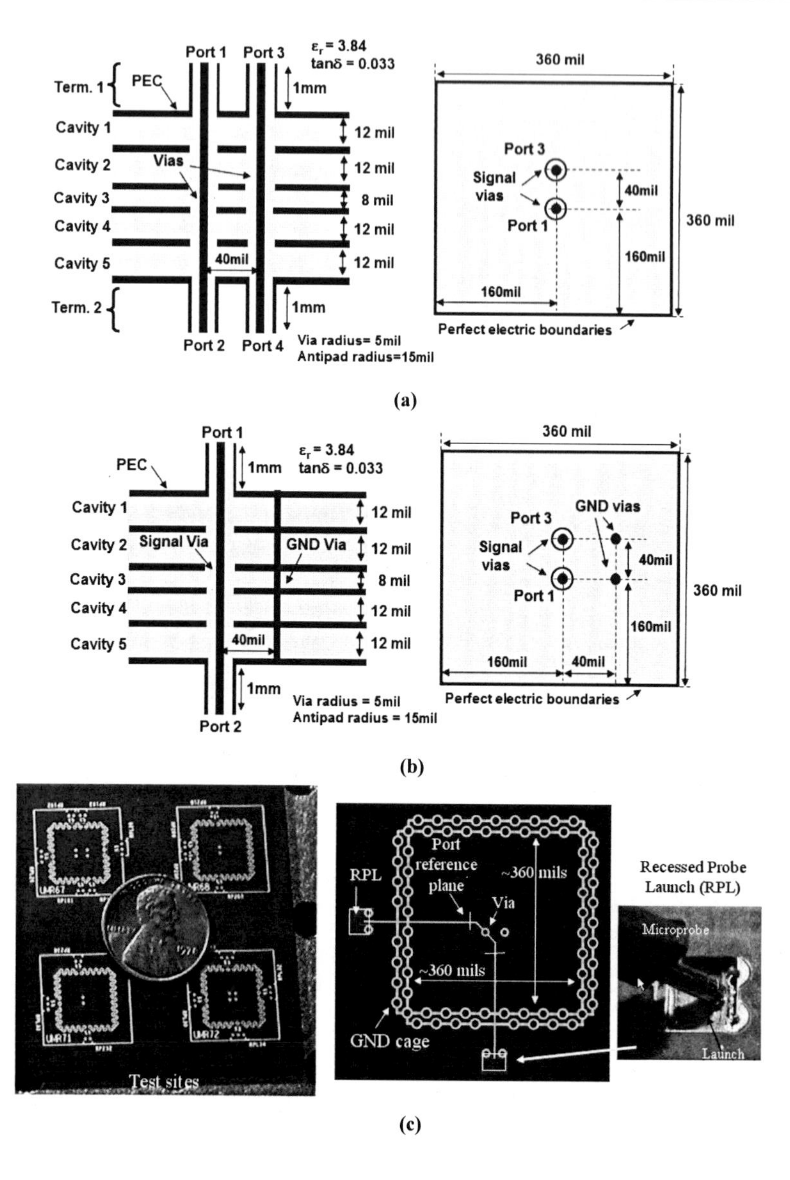

Figure 5.1 Multilayer via modeling examples. (a) Test structure with two signal vias. (b) Modified structure with two signal vias and two ground vias. (c) Test sites: picture, top layout view and detail of the recessed probe launch (RPL) [128]. The structures are enclosed by perfect electric boundary condition (PEC) and all the 1-mil-thick reference planes are assigned to ground. Computed via-to-plane capacitance values C^v for each cavity side were 29.1 fF (12 mil cavity) and 20.9 fF (8 mil cavity). *Photos courtesy of IBM Research.*

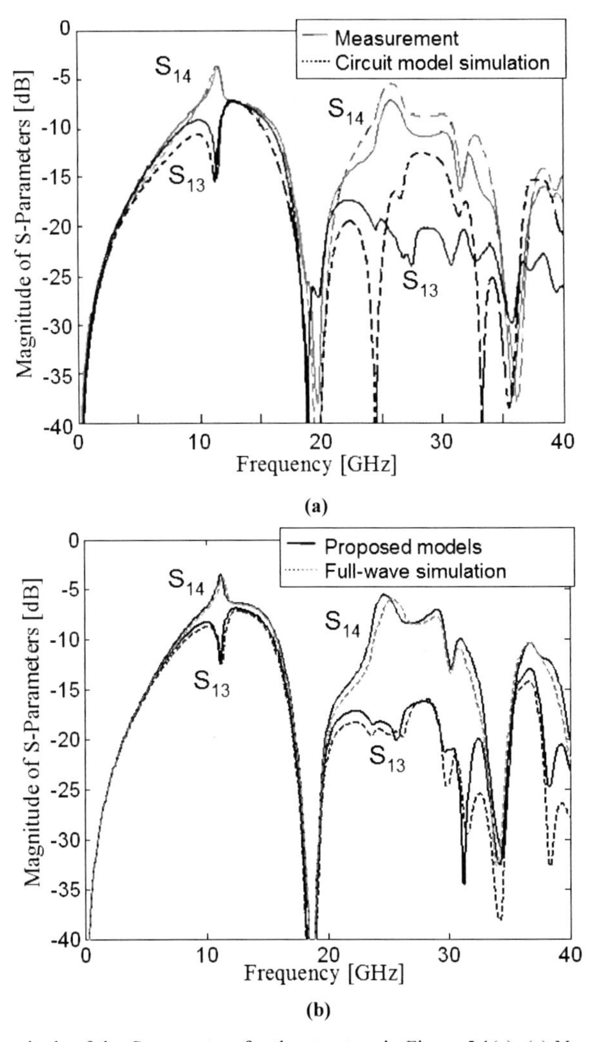

Figure 5.2 Magnitude of the *S*-parameters for the structure in Figure 5.1(a). (a) Near-end (S_{13}) and far-end (S_{14}) crosstalk reported in [94], obtained with the circuit-based approach and measurement. (b) Near-end (S_{13}) and far-end (S_{14}) crosstalk, obtained with the implemented code (model formulation in microwave network parameters) and full-wave simulation.

the circuit-based implementation of the models reported in [94] are also consistent. The deviations observed beyond 20 GHz were expected since the frequency dependencies of material coefficients were not considered; the recessed signal launches [26],[128] used to probe the physical structure and the microstrip-to-via transitions were also not modeled.

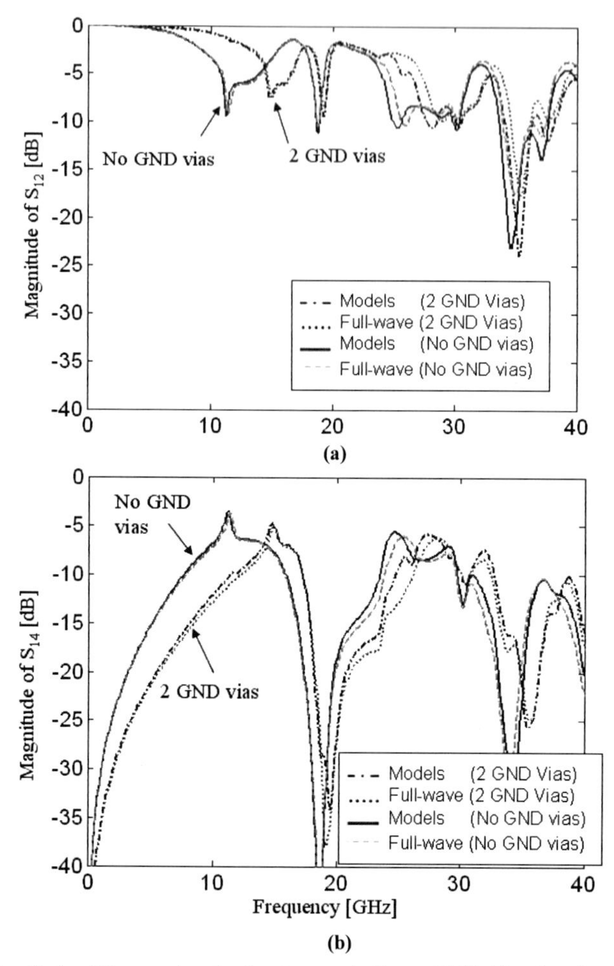

Figure 5.3 Magnitude of *S*-parameters for the structure in Figure 5.1(b). The plots show the effect of the ground vias, captured by the physics-based solution and the 3D FEM simulation. Transmission (S_{12}) and far-end crosstalk (S_{14}) are included.

An additional structure was simulated to evaluate the effect of return vias. Figure 5.1(b) shows the modified array with four vias, two signal and two ground vias. The results are included in Figure 5.3. The solution with the models was able to capture the influence of the ground vias, showing a very good correlation up to 20 GHz with respect to full-wave simulations. The main features of the response could also be reproduced for the rest of the evaluated frequency range. The computation times for this case are displayed in Table 5.1, considering different approaches to concatenate the partial

Table 5.1 Computation times for the example in Figure 5.1(b) *

FEM full-wave simulation (500 freqs)	Proposed models (500 frequency points)		
	ABCD matrix multiplication for result concatenation	Segmentation for result concatenation, reducing ground vias by Schur complement	Segmentation for result concatenation, without previous reduction of ground vias
5 h 32 min	14 s	19 s	35 s

*computed on a 32-bit 2.4-GHz CPU with 2 GB RAM.

cavity results. The full-wave simulation took several hours (elapsed time, 8 adaptive passes, no symmetry). In contrast, the analysis with the proposed models required less than 35 seconds under pessimistic conditions (no reciprocity and double summation to compute Z^{pp}). For these solutions the same workstation with 500 frequency points was used, and 40x40 modes were taken for the computation of the parallel-plate impedance.

5.2. Single-Ended Links

The validation of the models has been carried out by evaluating many different link configurations, considering different via locations, trace routing and lengths, offset striplines, as well as special types of vias. The examples in this section represent a selection of the most relevant studies concerning single-ended links.

5.2.1. Validation of the Via and Trace Model

A simple scenario was considered to validate the via and trace model, as described in Figure 5.4. The trace is centered (k = -0.5), it has a characteristic impedance of approximately 52 Ω and is 600 mil long. The extended model for vias and traces, according to Eq. (4.38), is applied to calculate the first cavity, whereas the other cavities are computed using Eq. (4.4). The results can be combined either by using *ABCD* matrix multiplication or segmentation techniques. Two coaxial extensions, 1 mm long, are used to place the external ports on the upper side. This was done to emulate external transmission lines contacting vias and to avoid placing the excitation too close to the beginning of the via discontinuity in full-wave simulations. The lower via ends were left open.

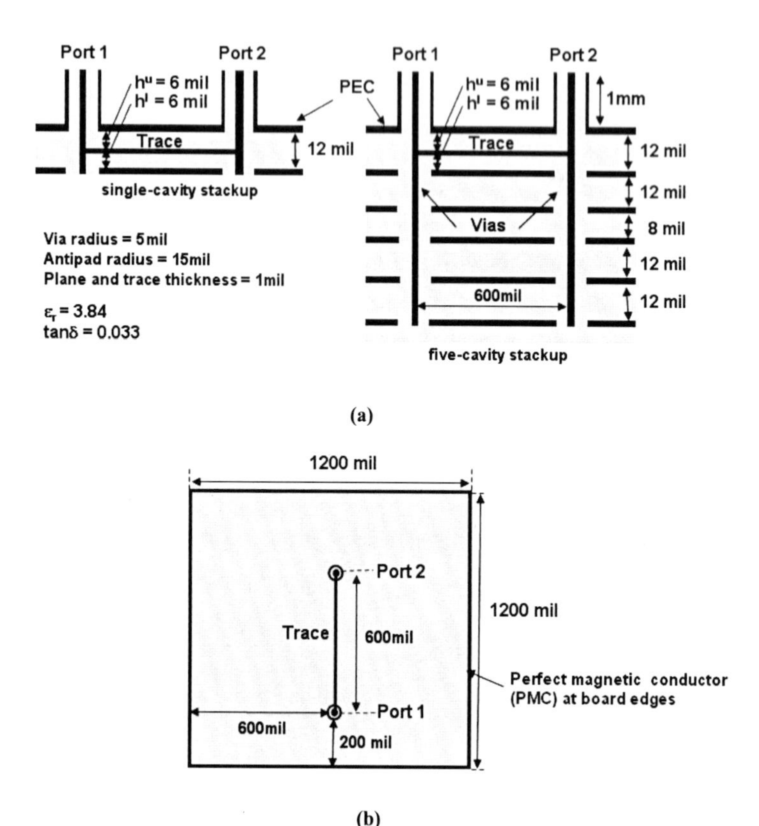

Figure 5.4 Structure for validation of the via and trace model. (a) Simulated cross sections with one and five cavities. (b) Top view of the structure detailing via and trace locations. Computed via-to-plane capacitances C^v per cavity side were 29.1 fF (12 mil cavity) and 20.9 (8 mil cavity).

Two cases were considered, including the simulation of a single cavity and the simulation of a five-cavity stackup with the trace located in the first upper cavity, where the via stubs are noticeably longer. The results obtained by the semi-analytical models and full-wave simulations are shown in Figure 5.5 and Figure 5.6, defining PMC boundaries at the board edges in the two methods. Both analyses are able to capture the via stub effect and predict worse transmission for the five-layer case. The agreement lies within 2 dB for almost all frequencies up to 20 GHz and it is still fair at higher frequencies. Deviations with respect to the full-wave solution tend to increase beyond 20 GHz. The differences observed can be consequence of the analytical assumptions, e.g.

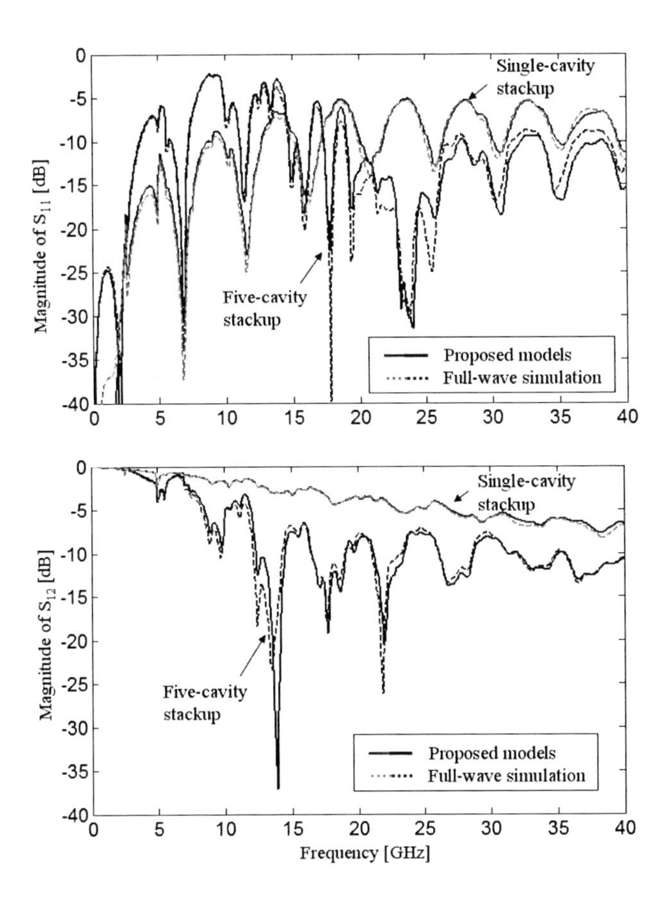

Figure 5.5 Magnitude of the *S*-parameters for the evaluation case of two vias connected by a trace (Figure 5.4), considering a single-layer and a five-cavity stackup. The results were computed by the proposed models and FEM full-wave simulations, assuming ideal PMC boundaries at the board edges.

symmetric field approximation around vias, or model parameter estimations. The different methods may also predict slightly different trace impedance values. Moreover, the lower via ends are assumed to be ideal open circuits, thus no radiation coming from these apertures was taken into account.

In general, it has been observed that the agreement of the solution with respect to full-wave simulations tends to be degraded for short vias connected by traces, as shown in Figure 5.5. This is attributed to the field distortion associated with the open ends

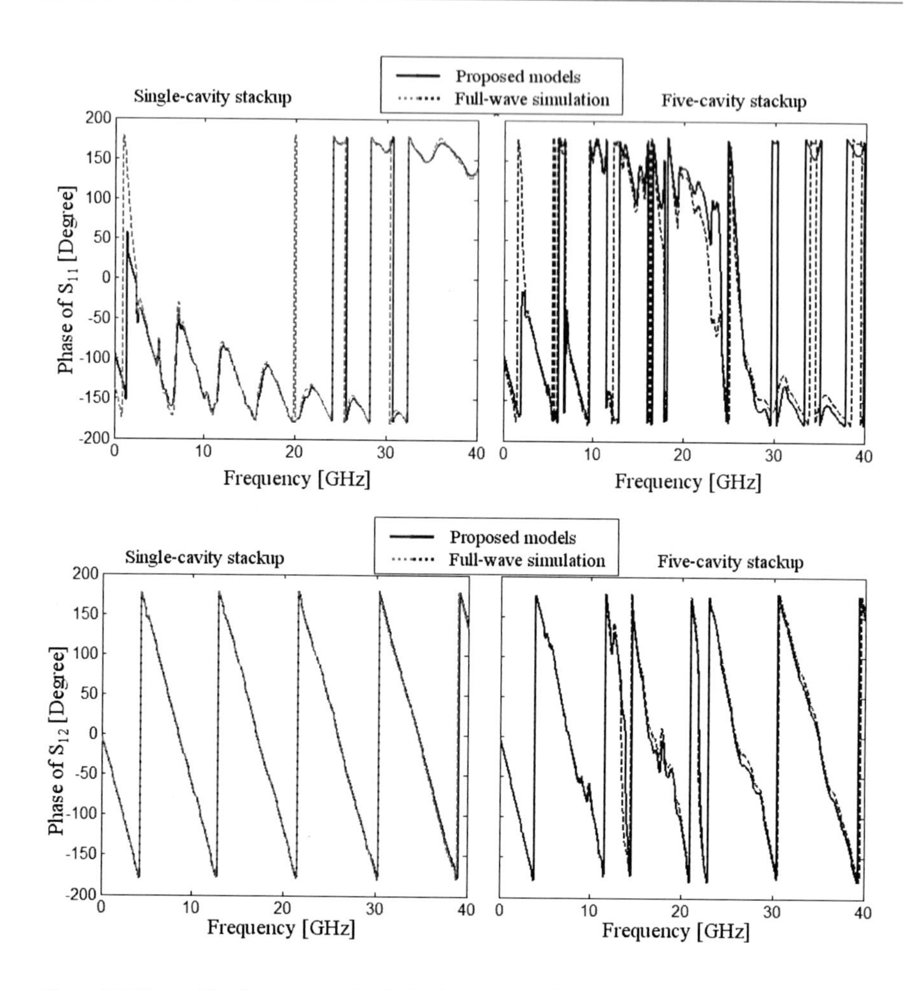

Figure 5.6 Phase of the S-parameters for the evaluation case of two vias connected by a trace (Figure 5.4), considering a single layer and the five-cavity stackup. The results were computed by the proposed models and FEM full-wave simulations, assuming PMC boundaries at board edges.

and the via-trace transition, whose impact becomes more pronounced for short structures and it is not rigorously modeled.

Regarding the computation time, for the five-cavity case the full-wave simulation took over 5 hours (FEM, 400 frequency points, no symmetry, single 2.4-GHz CPU, 2-GB RAM); the semi-analytical solution could be calculated in only 13 seconds (Matlab, single 2.4-GHz CPU, 2-GB RAM) using the double summation with 100 x 100 modes for the Z^{pp} calculation, assuming reciprocity and 400 frequency points.

5.2.2. Simulation of Single-Ended Links and Correlation to Full-Wave Simulations

The simulation of single-ended links is further analyzed with the structure described in Figure 5.7. Two links formed by single-ended striplines routed at different signal levels are connected by through-hole vias and a central turn via, used for the transition of the trace level, in a six-cavity square board with floating planes. The ports were defined at the top of the two vias, whereas all the other via ends were left open. The transmission line models were obtained from closed-form expressions for striplines [53] assuming a homogeneous and constant cross section. The cavity resonator model was used to compute the parallel-plate impedance and the via-to-plane capacitances were calculated analytically by Eq. (4.18). The obtained capacitance value was approximately 31 fF per cavity side. The uppermost and lowermost cavity sides have a slightly larger capacitance value of about 33.5 fF since the coaxial section for C^c in Eq. (4.18) is not bisected into two regions as is the case for the inner cavities. The fringing capacitance at the via ends was neglected.

The frequency and time domain results are shown in Figure 5.8 and Figure 5.9, respectively. They were compared with the results gained by full-wave commercial tools such as the finite integration technique in the time domain [83] and the finite element method in the frequency domain [127]. Conversion from frequency to time domain and vice versa was done by fast Fourier transform (FFT and IFFT, correspondingly).

Good agreement could be observed for the S-parameters results both in magnitude and phase. The feature selective validation technique [129]-[132] has been applied to quantitatively compare the results obtained with the different methods. Metrics from excellent to good were obtained for the magnitude of the S-parameters up to 20 GHz and between excellent and fair up to 40 GHz [19]. The discrepancies for the reflections at lower frequencies are attributed to numerical issues predicting relatively small values, given that the meshes and the solutions are adapted for a much larger frequency in the full-wave analyses. This region is also highly sensitive to trace impedance and via-to-plane capacitance variations in the models.

The computation times are summarized in Table 5.2. The proposed models provide accurate results and drastically reduce the computation time by about two orders of magnitude –in a conservative estimate– when compared to full-wave methods. In addition, acceleration methods to compute the parallel-plane impedance were used to improve the overall efficiency. For the case studied, the simulation time could be reduced by more than 50 % when the parallel-plate impedance was calculated with a single summation formula [105] instead of the double summation (Eqs.(4.5) and (4.10)).

Some approaches in the literature have addressed the modeling of power/ground structures by considering only the parallel-plate impedance and neglecting the near

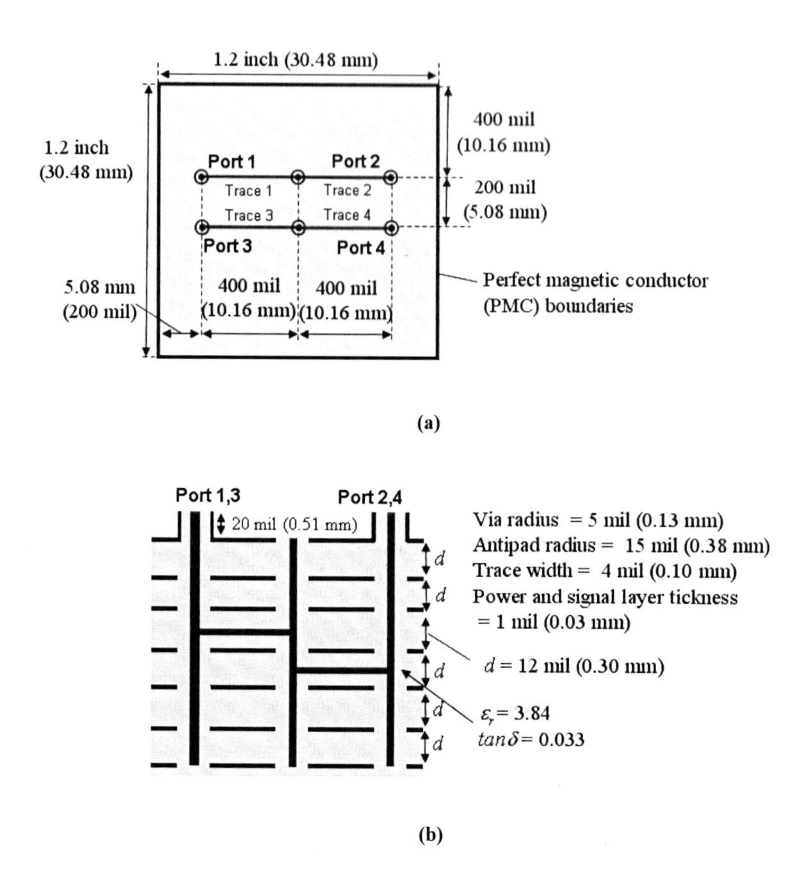

Figure 5.7 Test case used to validate the model against full-wave methods. (a) Top view of the structure. (b) Cross section of the board stackup.

fields in the antipad region. This procedure may be accurate for single layers and/or relatively low frequencies [103]. Figure 5.10 indicates that the contribution of the via-to-plane capacitances, used to approximate the near-field region in the models, is quite significant above 5 GHz in the current example. These capacitances are therefore important elements to consider for the modeling of multilayer boards.

Several other configurations that can be commonly found in real board and package designs have been modeled. The next subsections describe some test cases which are derived from the baseline structure of Figure 5.7.

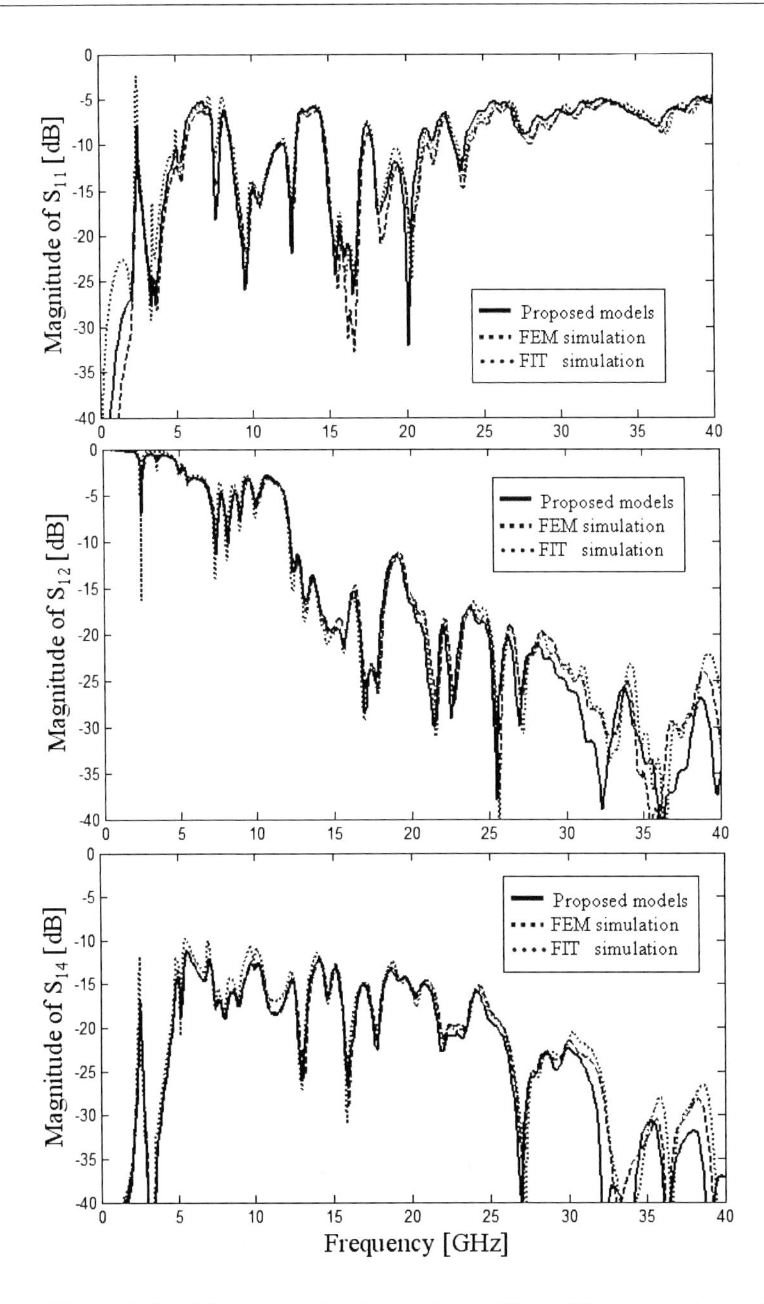

Figure 5.8 Magnitude of *S*-parameters for the structure in Figure 5.7, computed by the finite integration method (FIT), the finite element method (FEM), and the proposed models: reflection (S_{11}), transmission (S_{12}) and far-end crosstalk (S_{14}).

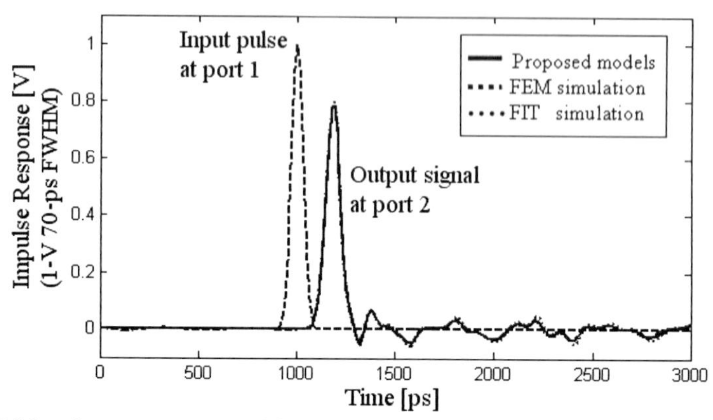

Figure 5.9 Impulse response computed for the structure in Figure 5.7. A 1-V 70-ps full-width half-maximum (FWHM) pulse is applied to port 1.

TABLE 5.2. Computation times obtained by different methods for the case in Figure 5.7*

Finite element method simulation (200 frequencies)	Finite integration technique simulation	Proposed models (200 frequency points)	
		Cavity model, double sum, 100x100 modes	Cavity model, single sum, 50 iterations
11 144 s (~3 h 5 min)	24 804 s (~4 h 53 min)	23 s	9 s

*computed on a 32-bit PC 3.0-GHz CPU, with 4 GB RAM.

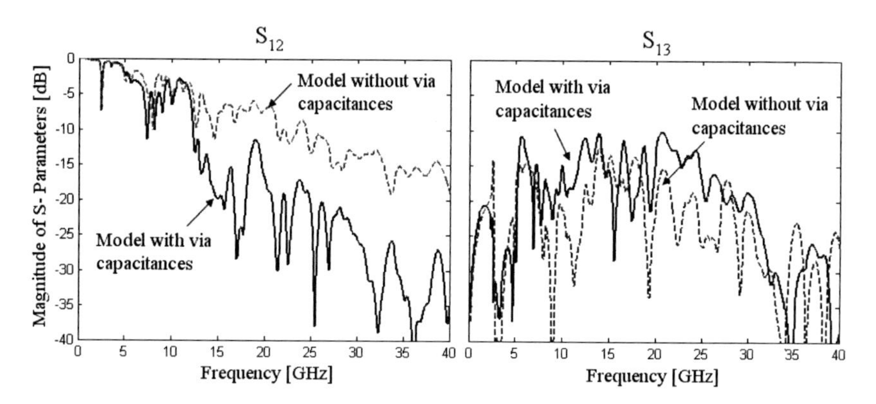

Figure 5.10 Contribution of via-to-plane capacitances to the total model results, computed for the structure in Figure 5.7.

5.2.3. Effect of Ground Vias

It is well known that ground vias can improve the electrical performance of interconnects by providing additional return paths to the signals traveling along signal vias [16],[87]. As mentioned before, the proposed method is capable of modeling structures containing both signal and power/ground vias by accounting for their connectivity with the reference planes that are used for ground or for power distribution. Return vias connecting both top and bottom planes of a cavity can be reduced right after the computation of the parallel-plate impedance by applying the Schur's complement (Eq. (4.22)). This operation improves the overall efficiency of the computation since it reduces in an early stage the size of the matrices [16]. Signal to ground via ratios (S:G) of 1:1 and 1:2 were studied for the configuration in Figure 5.11, assuming that all reference planes correspond to ground planes (with the stackup in Figure 5.7). The models were able to capture the effect of ground vias, showing good agreement with respect to full-wave methods. In Figure 5.11(b), it can be observed that the near-end crosstalk is mitigated as the number of ground vias is increased. The plot also includes the comparative results obtained by FEM simulations. The effectiveness of ground vias to improve the transmission and to reduce the crosstalk is better for lower frequencies. In this case study, above 25 GHz it is difficult to differentiate between the three cases, as the predicted crosstalk levels become similar for all of them. As the frequency increases more ground vias located in close proximity to the signal via become necessary to obtain an appreciable crosstalk reduction.

5.2.4. Mixed Reference Power/Ground Planes

In contrast to the cases discussed before, practical designs usually contain interleaved power and ground reference planes. The models and the proposed method are flexible enough to handle these scenarios by properly interpreting the connectivity of the power vias with respect to the reference planes (see Section 4.5). This is an important advantage of the method, since it allows the modeling of structures with realistic complexity and arbitrary stackup definitions.

The impact of interleaved ground and power planes was investigated and compared with the configuration where all the reference planes are assigned as ground. In this example, the mixed reference planes consider all the inner planes as floating power planes and the top and bottom planes as ground (Figure 5.12(a)). The models were able to capture with good accuracy the subtle differences in the response predicted by a full-wave simulation (Figure 5.12(b)).

Despite the number of floating planes, the difference in the S-parameters between the two cases is relatively small. The return path provided by ground vias is present even if the via does not touch some inner reference planes. The additional coupling

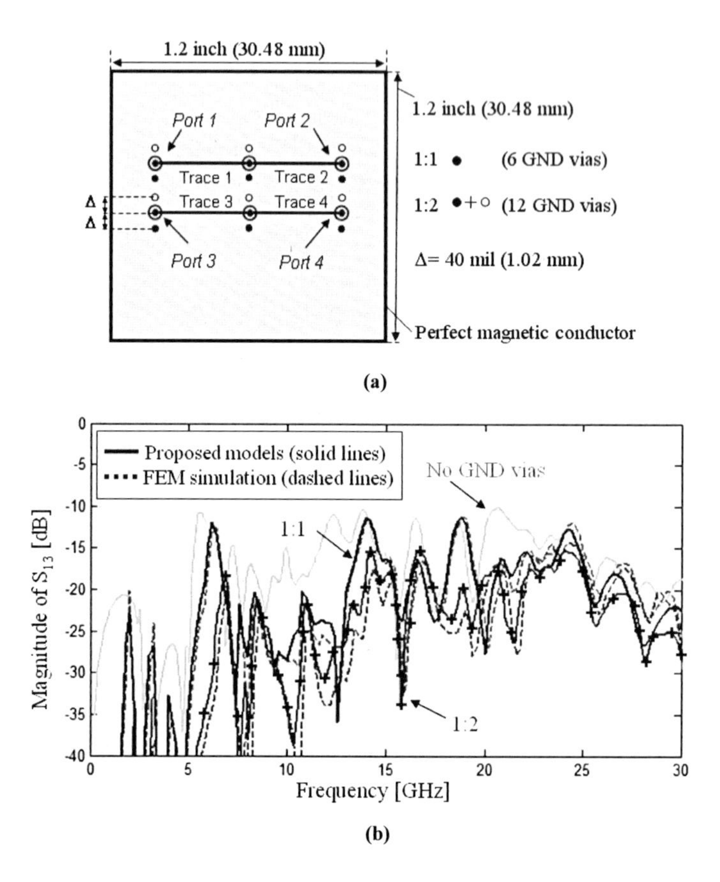

Figure 5.11 Test cases with ground vias. (a) Structure top view. (b) Near-end crosstalk for different ground to signal via ratios, obtained by the models and finite element method (FEM) full-wave simulations. The stackup corresponds to the one shown in Figure 5.7, with all the reference planes defined as ground.

paths between cavities provided by the clearance holes on planes (antipads) in the mixed reference case may explain the slight differences observed and the trend towards lower crosstalk for the case where all the planes are defined as ground. This tendency depends on the frequency, and it holds better for the lower frequency range. At higher frequencies it becomes more difficult to perform a trend analysis due to the rapid variations of the system response. Alternative reference plane combinations were studied using the structure of Figure 5.7. As expected, when more intermediate ground planes were added, the differences diminished in comparison to the case with all reference planes defined as ground.

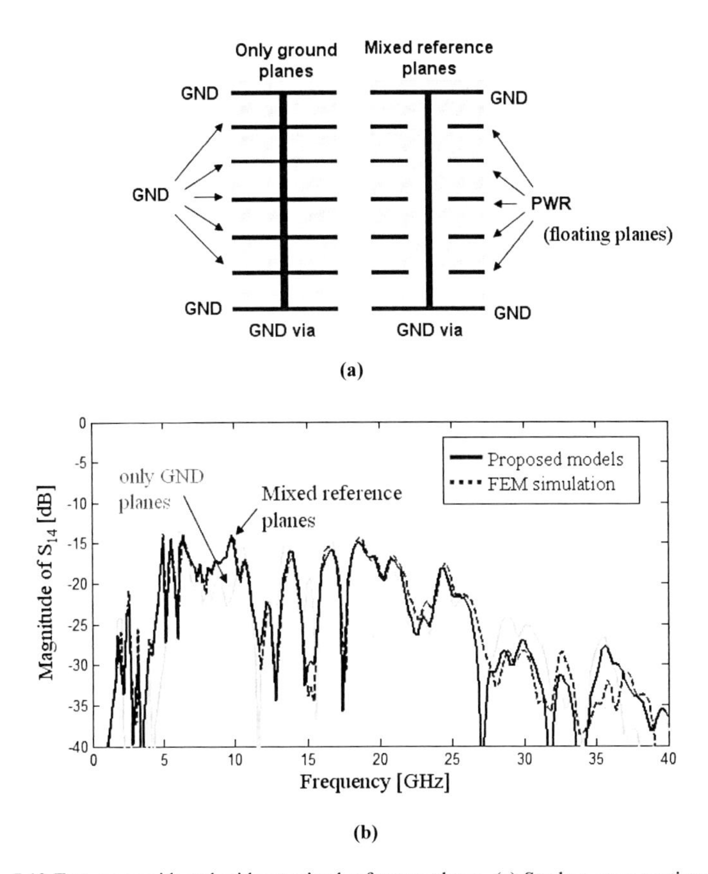

Figure 5.12 Test cases with and without mixed reference planes. (a) Stackup cross-sectional views. (b) Far-end crosstalk for both cases and correlation to a finite element method (FEM) full-wave simulation.

Recently, the role of mixed reference planes on single-ended and differential links has been further studied in [8]. The results also confirm that the best link performance is achieved if the number of reference planes defined as ground is increased, since a larger number of ground planes provides, in principle, a better return path for the signal current. Insertion loss, but in particular crosstalk and mode conversion tend to increase if more power planes are used. At single frequencies the plane assignment may accentuate or shift plate resonances. The effect on transmission can be narrowband or more broadband depending on the environment of the link, including for instance position and type of surrounding vias, size of the reference planes, etc..

5.2.5. Blind and Buried Vias

Another case of interest that can be addressed with the models proposed in this work is the simulation of non-through-hole vias that only partially traverse the board cross-section, e.g. blind and buried vias. The investigations carried out have shown that it is possible to simulate structures containing blind vias and/or buried vias crossing about one half of the cavity with simple approximations for the near fields at the via transition.

For these cases, the via model has been applied and only the via-to-plane capacitances for the internal end segments of the via are recalculated. This means that the parallel-plate model has not been modified for buried vias. This idea is illustrated in Figure 5.13, where the capacitances C^{vb} are the parameter to be extracted for a blind or a buried via. This capacitance value is calculated with quasi-static solvers. The test structure discussed corresponds to the case in Figure 5.7 replacing the through-hole vias with blind and buried vias. Figure 5.14 shows that, with this approximation, good results are achieved when handling these special types of vias up to 20 GHz.

A limitation of this approach is that the model will not be able to consider the effect on the parallel-plate modes due to a buried via transition, which just penetrates a section of the cavity. This may start to become important as the frequency increases and could explain the discrepancy observed at frequencies beyond 20 GHz between the results predicted by the models and the full-wave simulation.

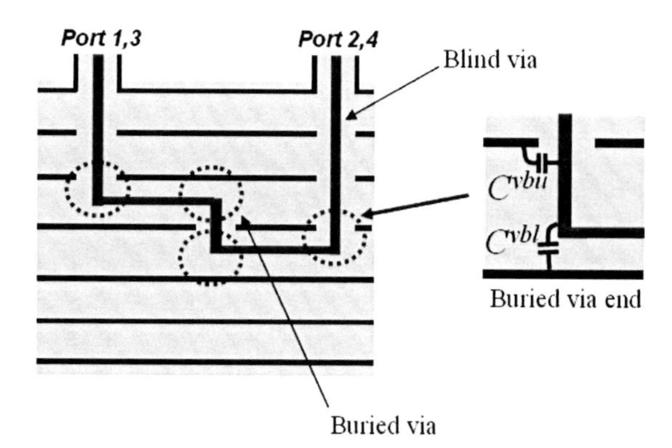

Figure 5.13 Cross section diagram of the modified test structure, replacing through-hole signal vias in the structure of Figure 5.7 with blind or buried vias.

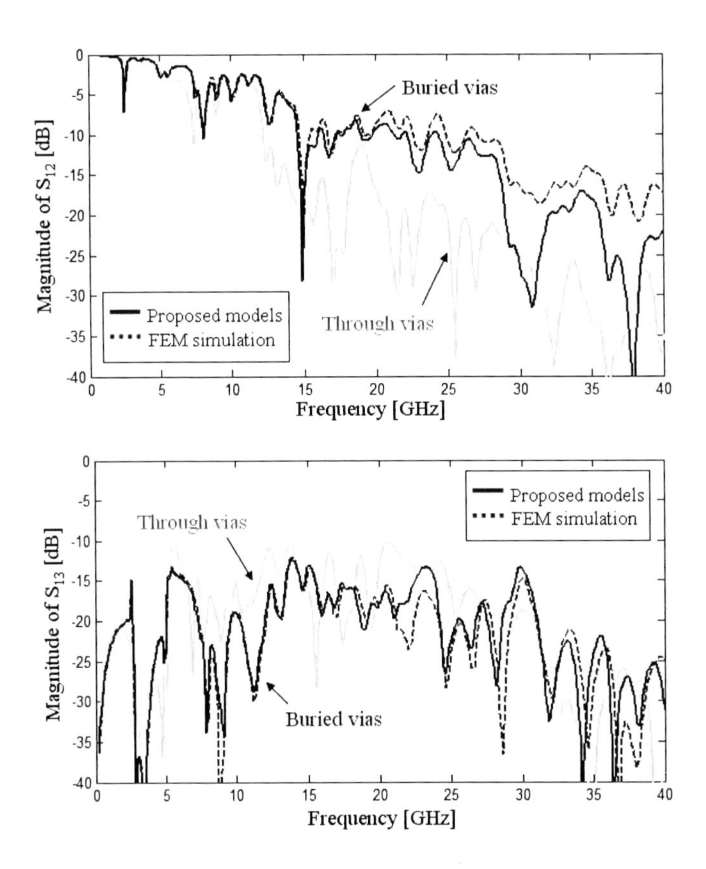

Figure 5.14 Transmission and near-end crosstalk for the configuration with blind and buried vias in Figure 5.13, obtained with the models and finite element method (FEM) full-wave simulations. The results are compared also with the ones obtained for the case with through vias (Figure 5.7).

5.2.6. Model-to-Hardware Correlation

The models have also been validated against measurements on PCB test structures. The configurations are described in Figure 5.15, which are comparable to the cases previously analyzed. The measurements were performed in the frequency domain, with a 4-port vector network analyzer (VNA) and ground-signal (GS/ SG) microprobes of 225 μm pitch, placed on the surface of the board. A SOLT –Short-Open-Load-Thru standard based– calibration was performed up to the probe tip on an impedance-

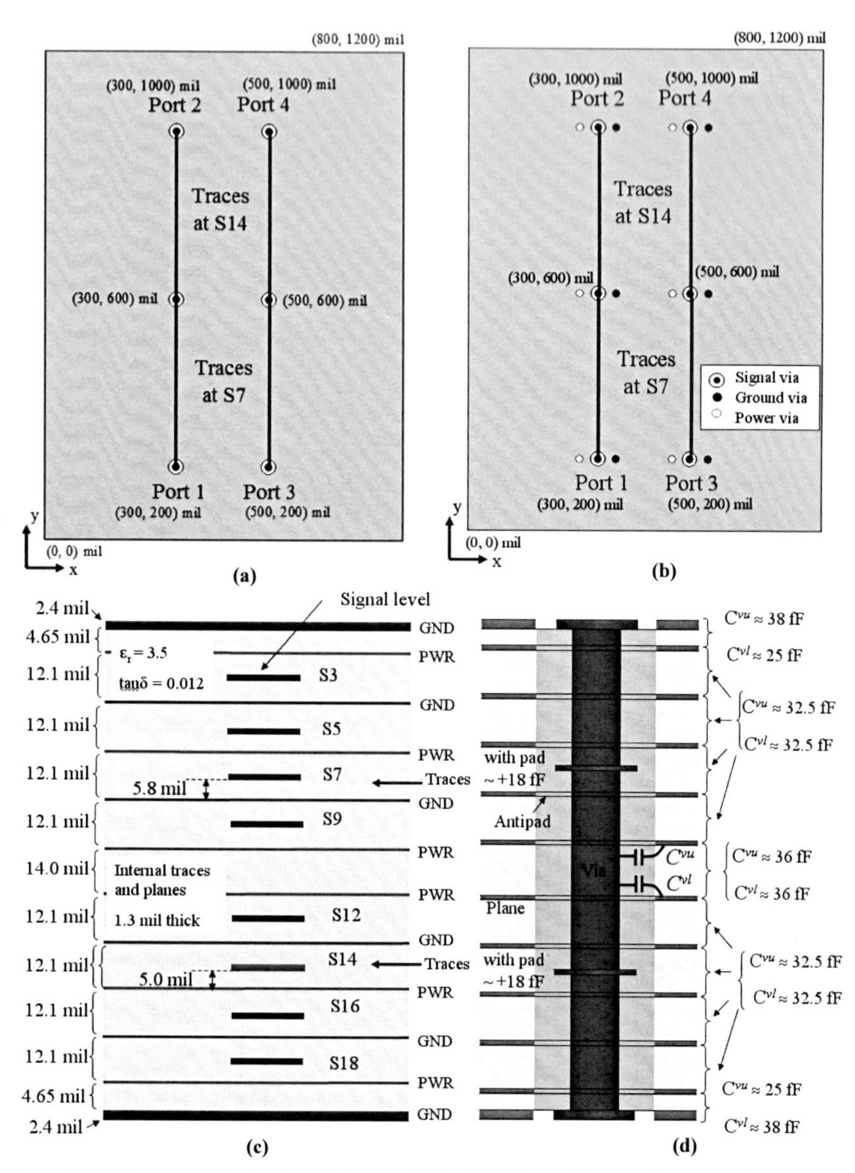

Figure 5.15 Diagram of the single-ended test vehicles (TV) for model-to-hardware correlation. The vias are through-hole and have a radius of 6.9 mil, the pad radius is 11.9 mil, the antipad radius is 17.9 mil, and the traces are 5.25 mil wide. (a) TV-1. (b) TV-2. (c) Stackup definition. (d) Via capacitance values extracted with a quasi-static solver [119]. If a pad exists at signal layers, the via-to-plane capacitance value increases approximately 18 fF for both top and bottom cavity sides.

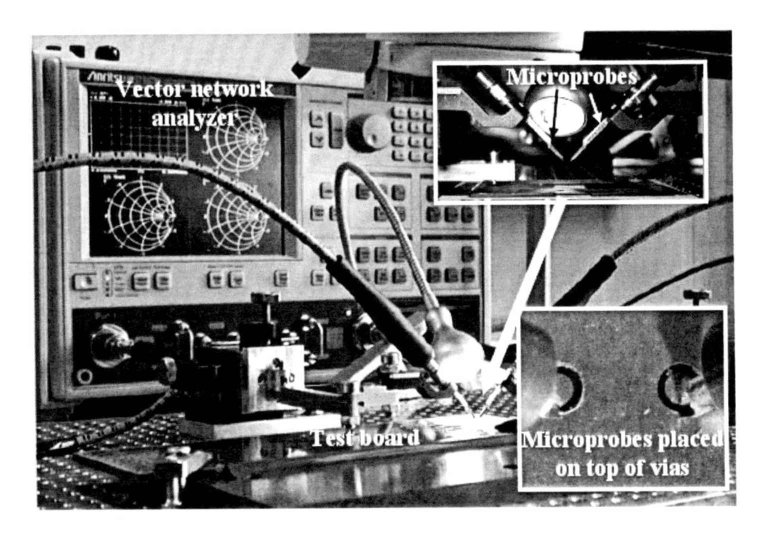

Figure 5.16 Typical measurement setup for PCB structures in the GHz range. The image shows a vector network analyzer, low-loss phase controlled cables and the microprobes used to contact the structure. The placement of the probes on top of the vias has been magnified. *Photos courtesy of Miroslav Kotzev at the Institute of Electromagnetic Theory (TET), TUHH.*

controlled substrate. Figure 5.16 depicts a similar measurement setup available at the Institute of Electromagnetic Theory, TUHH.

Two single ended links routed between through vias are considered, with 400-mil traces at S7 and S14 levels connected by a turn via. Figure 5.15 shows two configurations, with only signal vias (TV-1), and adding ground and power vias (TV-2). The power and ground vias are placed in a 1 mm pitch and they are contacting all the planes defined as power or ground, respectively. The stackup has eleven cavities and eight signal levels, from which only S7 and S14 are used (Figure 5.15(c)). The traces are not centered and they have k-factors of -0.53 and -0.47 for the S7 and S14 levels, respectively. The trace models were obtained from a FEM full-wave simulation. Constant material parameters are assumed for the dielectric (ε_r = 3.5, tanδ = 0.012) and Copper conductor (σ = 5.8 $\cdot 10^7$ S/m). Four ports are defined on top of the vias as indicated. The cavity model single summation (Eq. (4.5)) with 100 iterations was used to compute Z^{pp}; each cavity was calculated separately, using 200 frequency points. The via capacitances were extracted with a quasi-static solver, considering the fringing capacitance at via ends. Pads are present at every via stripline transition. The extracted values are defined between the via barrel and the planes, and in a post-processing step they are separated as C^{vu} and C^{vl} for each cavity side (Figure 5.15(d)).

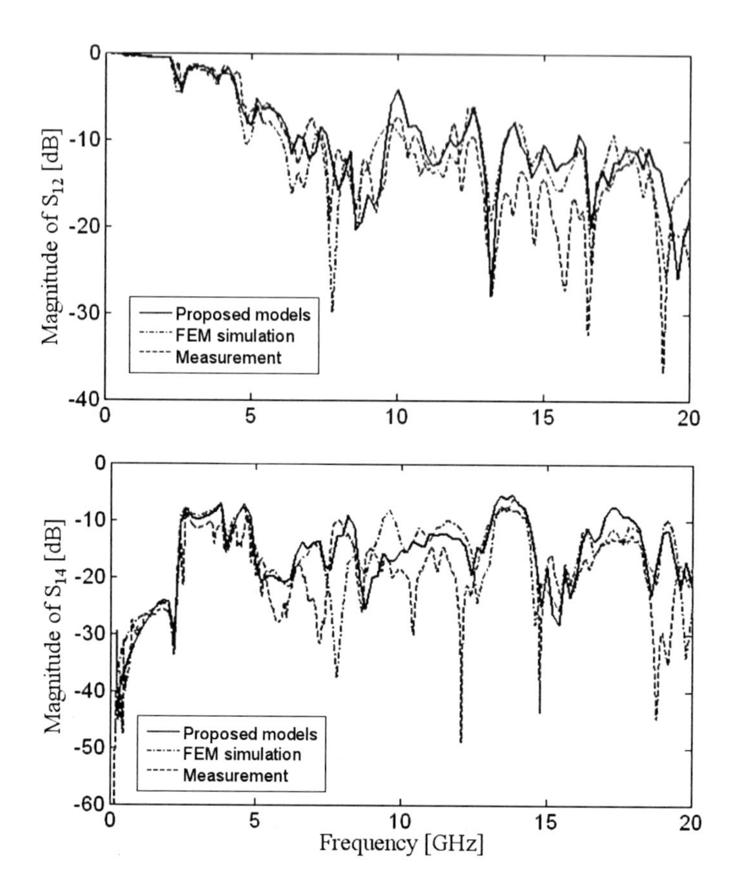

Figure 5.17 Magnitude of *S*-parameters for the TV-1 in Figure 5.15, obtained by measurement, FEM full-wave simulation, and the models.

The analytical calculation of the via capacitance for the layers without pads predicts values about 10 % larger in comparison to the ones obtained by the quasi-static solver. Due to convergence issues during the extraction, the capacitance values might vary by several fF and therefore it is estimated that they are accurate within 10 % for the inner layers and only up to 20 % for the fringing capacitances at via ends. These variations do not impact significantly the correlation in terms of *S*-parameters, however further research must be conducted to determine accurate procedures to extract the capacitances and to approximate the fringing fields in presence of pads and traces.

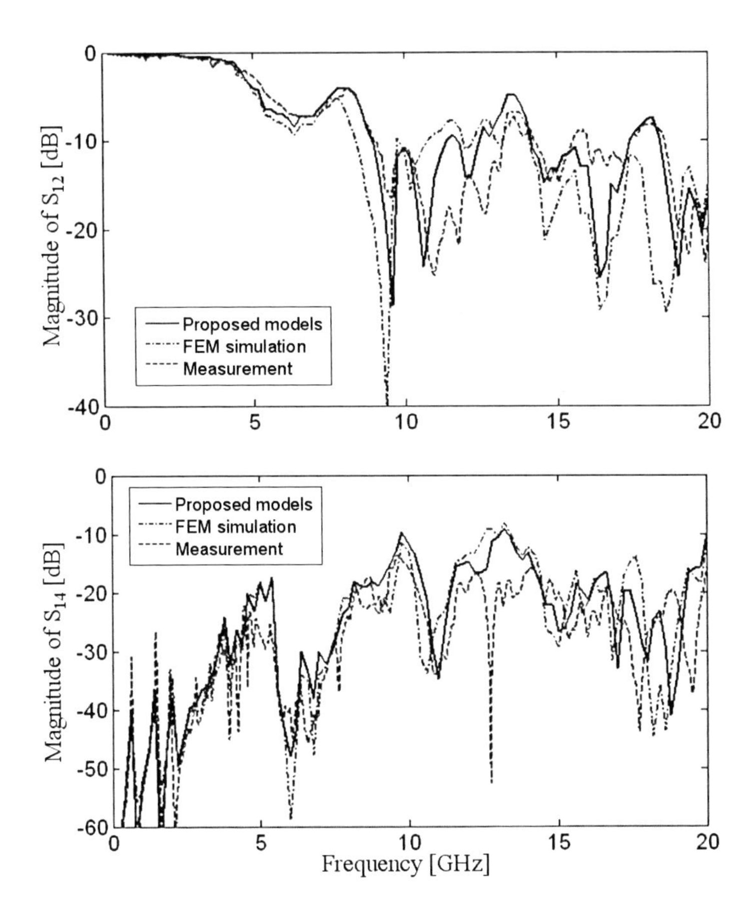

Figure 5.18 Magnitude of *S*-parameters for the TV-2 in Figure 5.15, obtained by measurement, FEM full-wave simulation, and the models.

The transmission and far-end crosstalk are compared for TV-1 and TV-2 in Figure 5.17 and Figure 5.18, using data from three different sources: the simulation with the proposed models, hardware measurement, and a FEM full-wave analysis. For a better visualization, the plots show the curves up to 20 GHz, due to the multiple resonances. The *S*-parameters up to 40 GHz can be found in Appendix C.1. The agreement obtained for the results of both structures is from fair to very good up to 20 GHz.

The three methods can capture the main characteristics of the response and predict similar levels. Note that the effect of power/ground vias could be detected in terms of crosstalk reduction and improvement of the transmission in the TV-2 solution when compared to the TV-1 case, in particular for the lower frequency range. Discrepancies in the vicinity of sharp resonances and the magnitude at some points still remain. The model and full-wave simulation results tend to agree better in comparison to the measurement. This was expected since in both modeling approaches the material and geometrical parameters are clearly defined and show no frequency dependencies. In practice and given the frequency range investigated, material coefficients are frequency dependent. The simulations also define perfect open boundaries at the board edges, whereas in the real structure the fringing fields at these locations, due to the number of cavities and the small size of the board, may start to acquire significance in the multi-GHz range. The real structure is subject to fabrication tolerances and process variations, and therefore the nominal design parameters used in the models are not exact. Moreover, the excitation is applied through the probe tips (point-wise), assuming that the calibration is good enough to remove the probe parasitics and neglecting any additional interaction among the probe and the board under test. For the models and full-wave analysis, circular lumped ports are utilized, assuming that the fields on them are isotropic and therefore constant in the axial φ direction. This assumption may also start to fail at higher frequencies. Further research is necessary to refine the extraction of model parameter and investigate the issues mentioned above.

The simulation with the proposed models took about 10 seconds and 50 seconds for TV-1 and TV-2, respectively, on a 3.3-GHz CPU with 4 GB RAM. These times do not consider the computation of the trace model and extraction of the via capacitances. The full-wave simulations took more than 20 hours running on the same computing platform.

5.3. Differential Links

The utilization of differential signaling is a common practice in high-speed designs due to the relatively high immunity to common-mode noise and reduced radiated emissions. The modal decomposition approach gives the possibility to model coupled conductors since the transmission line model can be computed independently from the parallel-plate model. Assuming that mode conversion only occurs at the via transitions, as discussed in Section 4.6, it is possible to generate the models using multiconductor transmission line (MTL) theory by means of any solver suitable for this purpose (analytical, 2D or 3D). The required output is the matrix for the single-ended MTL system with ports defined at the trace terminals, as illustrated in Figure 5.19 for differential striplines.

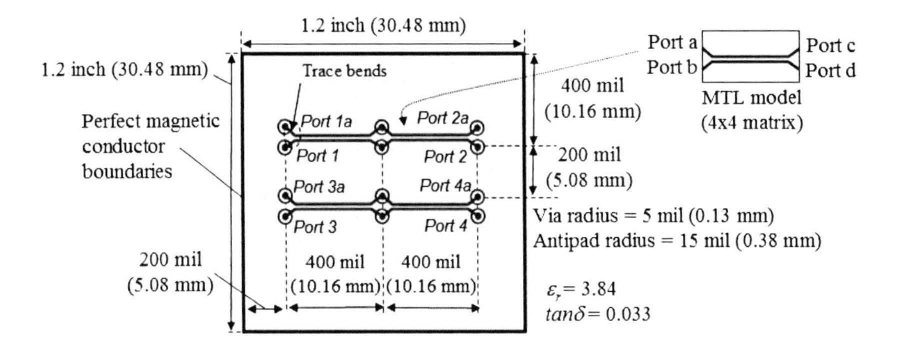

Figure 5.19 Test structure with differential links (coupled conductors). The sketch describes the top view of the structure, whose stackup is identical to the one in Figure 5.7.

5.3.1. Simulation of Coupled Striplines

For the validation of a case with coupled striplines, the example in Figure 5.7 was modified by replacing the single-ended vias and striplines with differential ones. The differential via ports are defined on top of the board, between adjacent vias labeled with the same number. The configuration in Figure 5.19 has twelve vias and two differential links routed on the third and fourth cavities. Figure 5.20 details the obtained results, expressed as mixed-mode S-parameters (Appendix A.8) [133]-[134], and compares them with the solution of a full-wave FEM analysis. The agreement for both cases is good and the improvement in the response is clear for the differential parameters in contrast to the common-mode ones, under ideal timing and symmetry conditions. The differential crosstalk for frequencies below 15 GHz is substantially lower in comparison to the common-mode case.

The transmission line model was obtained using a 3D FEM solver. Nonetheless, reasonable results can also be achieved with a 2D model derived from cross-sectional information that neglects the trace segments necessary to reach the vias, denoted as trace bends in Figure 5.21. Obviously, the portion of the trace with a non-constant cross section should be relatively short in comparison to the link length and minimum wavelength of interest.

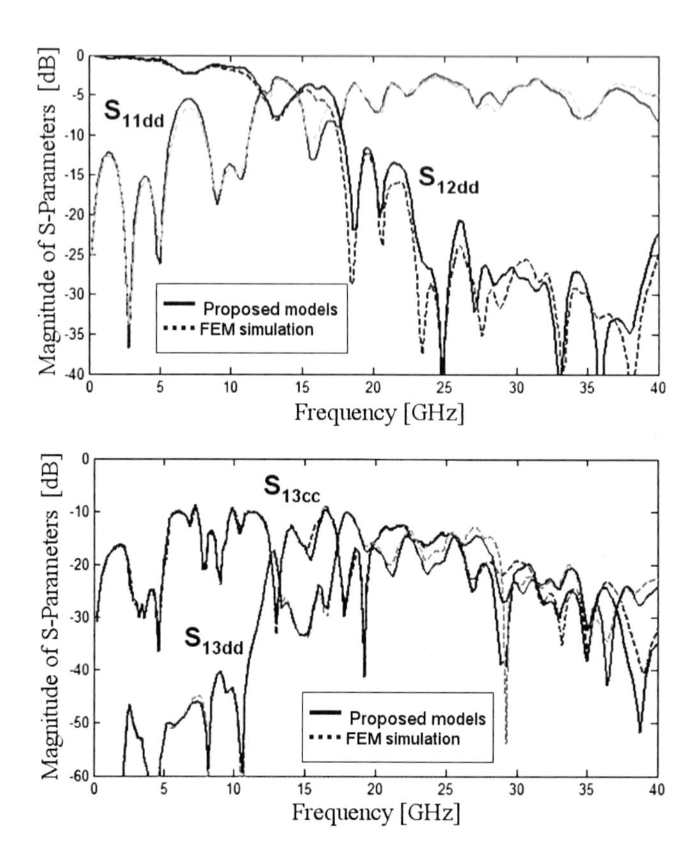

Figure 5.20 Magnitude of the mixed-mode *S*-parameters for the case in Figure 5.19. Differential reflection and transmission, and differential (*dd*) and common-mode (*cc*) near-end crosstalk are shown, as well as the correlation to finite element method (FEM) simulations.

5.3.2. Model-to-Hardware Correlation

The model-to-hardware correlation for a case with differential vias and traces is discussed with the test board TV-3 described in Figure 5.22. Four lumped ports are defined on top of the vias. A 0.8-inch differential link is routed on the S14 level with four thru vias at the trace terminations. Each signal via pair is surrounded by two ground and two power vias. The stackup and via parameters are the same as for the test vehicles in Figure 5.15. The measurements were done using the setup and procedure described in Section 5.2.6.

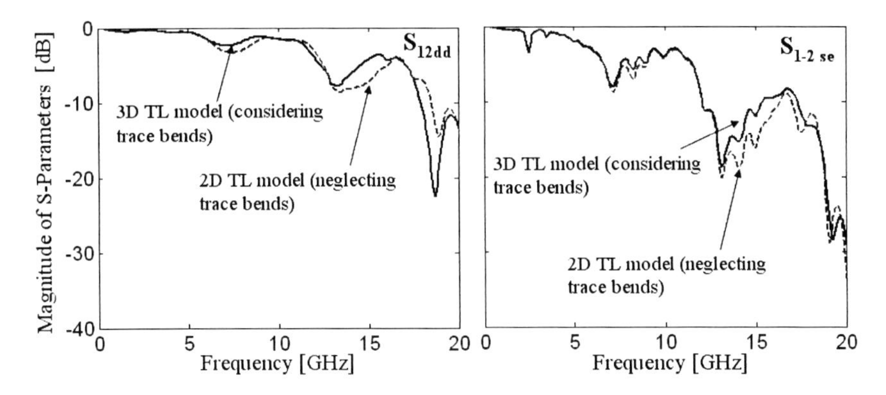

Figure 5.21 Effect of neglecting the trace bends on differential (*dd*) and single-ended transmission (*se*) between the ports 1 and 2, for the example of Figure 5.19 and using the proposed models.

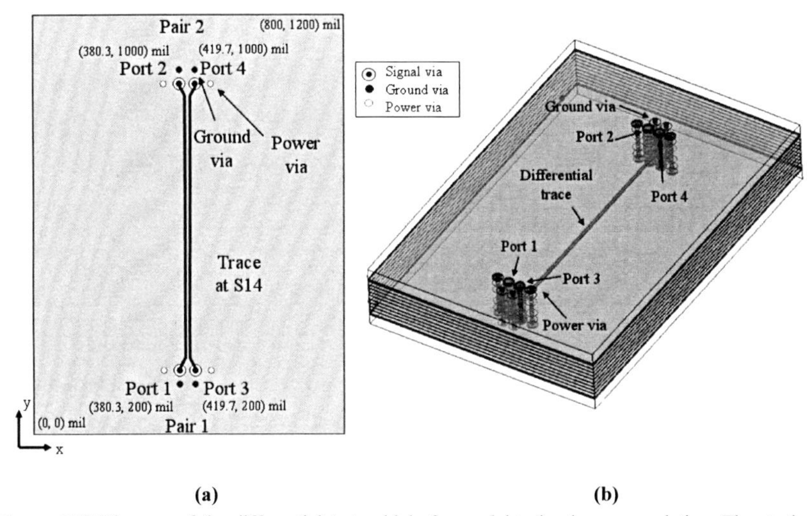

(a) **(b)**

Figure 5.22 Diagram of the differential test vehicle for model-to-hardware correlation. The stackup and geometric parameters are the same as defined in Figure 5.15. (a) TV-3. The adjacent vias are placed in a 1-mm pitch grid. (b) 3D full-wave model of TV-3.

For the analysis of the structure, the differential stripline model (4 ports) was obtained with a 3D FEM simulation that takes into account the trace bends. The cavity model single summation (Eq. (4.10)) with 100 iterations was used to compute Z^{pp}, calculating each cavity separately with 200 frequency points. The via-to-plane

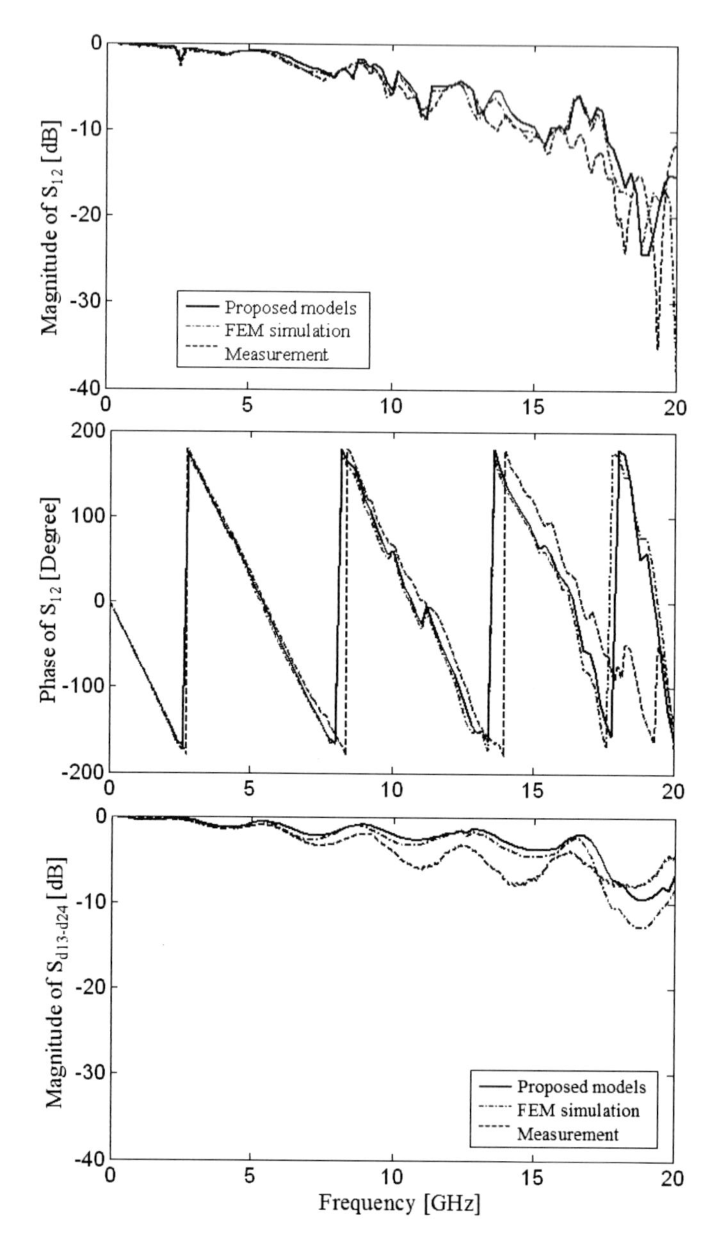

Figure 5.23 *S*-parameter single-ended and differential transmission for the TV-3 in Figure 5.22, obtained by measurement, FEM full-wave simulation, and the models.

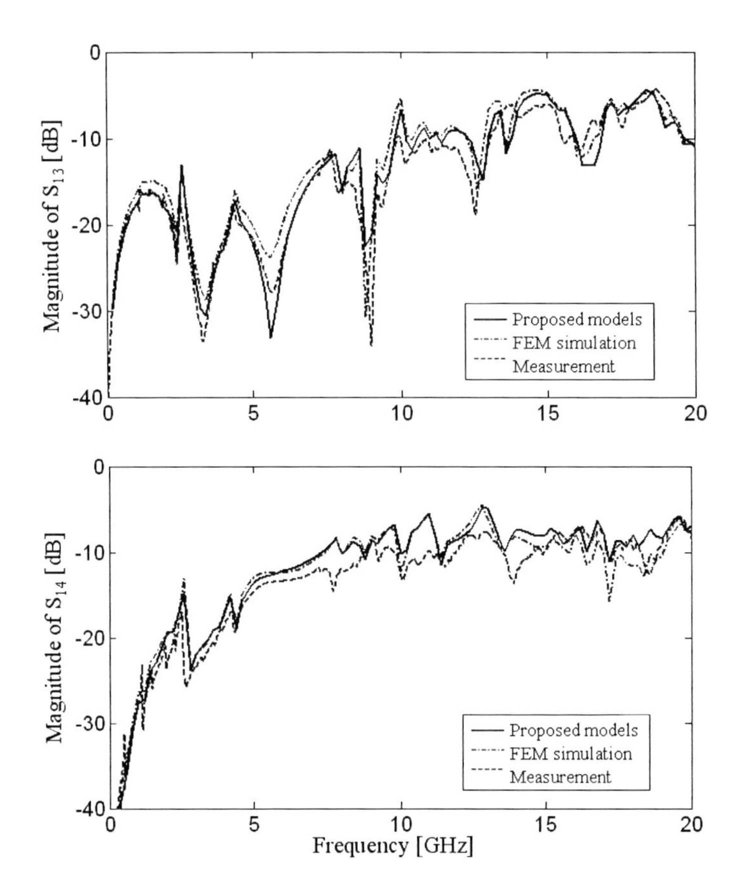

Figure 5.24 Single-ended crosstalk parameters for the TV-3 in Figure 5.22, obtained by measurement, FEM full-wave simulation, and the models.

capacitance values correspond to the ones provided in Figure 5.15, including pads at the S14 level where the traces contact the vias. The simulation time for this case was about 23 seconds (3.3-GHz PC, 4 GB RAM).

The comparisons between measurement, full-wave simulation, and the model *S*-parameters are provided in Figure 5.23 and Figure 5.24. The agreement is good up to 20 GHz considering the model simplifications, as discussed in Section 5.2.6. The results obtained with the models and the full-wave simulations agree well with each other. When compared to the measurement, the deviations observed are more significant. The phase of the single-ended transmission in the measurement indicates that after 8 GHz the difference starts to increase and the structure tends to look electrically shorter than predicted with the models and the full-wave analysis. The neglect of fringing fields at plane edges, deviations of geometrical parameters from the nominal values, and frequency dependencies of material parameters may be causing these differences. The mixed-mode parameters are in particular affected, where the simulated differential transmission is higher than the measurement up to 17 GHz. Figure 5.24 shows the correlation for the single-ended crosstalk parameters. The results up to 40 GHz are shown in Appendix C.2. Above 20 GHz the responses show similar characteristics, but the deviations tend to increase since the discussed correlation issues become more pronounced at higher frequencies.

5.4. Via Arrays

Vias are usually arranged into regular fields that contain power, ground and signal vias on a Cartesian grid. Ball (BGA) or Land (LGA) Grid Arrays are widely used packaging technologies which may comprise via fields with thousands of elements.

Test via arrays of size 8 x 8 (64 vias) have been designed and fabricated. An array with 29 signal and 35 ground vias, placed on an 80-mil grid is depicted in Figure 5.25. All vias have a radius of 5 mil, circular antipads with a radius of 15 mil, and circular pads at top and bottom via ends with a radius of 10 mil. The board has 18 metallic levels and 11 cavities. All reference planes are assigned as ground, including the flooded plane sections on top and bottom sides of the board. The dielectric material of the board is Nelco4000-13 [135], modeled with a dielectric constant of 3.7 and a loss tangent factor of 0.03.

The measurements were carried out at IBM and TUHH with similar hardware setups; a vector network analyzer and 225 μm pitch GS microprobes were used, calibrated on an impedance-controlled substrate with the SOLT procedure. The agreement observed between the two data sets indicates that the repeatability of the measurements is good. Two ports were defined on top of different vias per measurement, whereas all the other via ends were left open. Figure 5.25 also shows the two probes on top of the physical structures.

The real test vehicle is a larger panel shared by several test structures. The top and bottom reference planes are not continuous outside the array. For modeling purposes, it was assumed that the planes are infinitely large. This should serve as a reasonable approximation given the position of the defined ports, the number of ground vias, and the relatively large distance to other neighboring structures. The via-to-plane capacitances were extracted considering the fringing fields at via ends using a quasi-static solver.

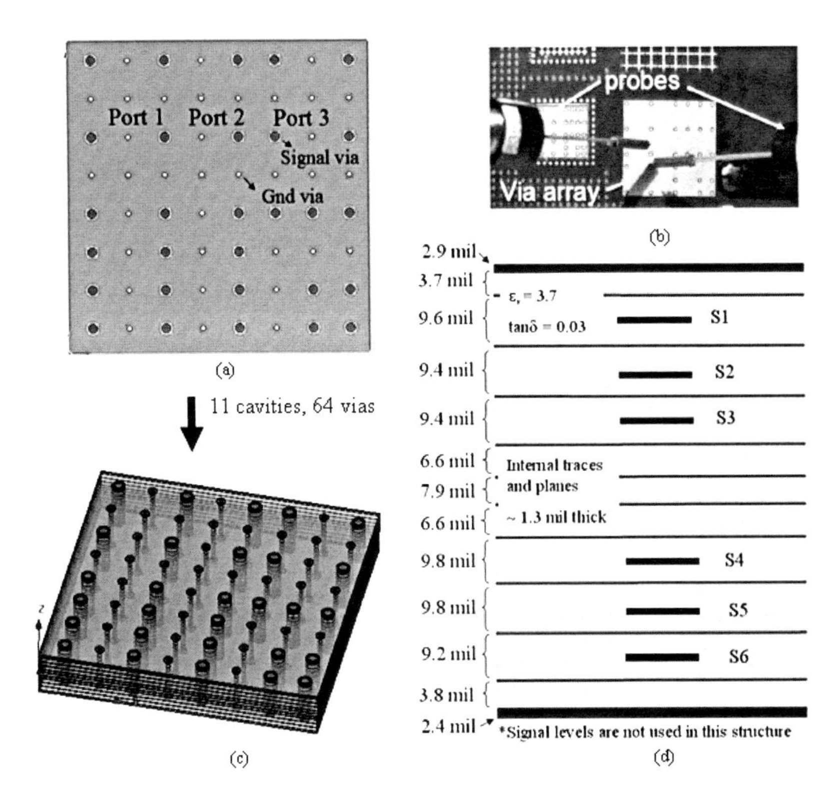

Figure 5.25 Description of the investigated via arrays. (a) Distribution of signal and ground vias, placed in a 80-mil pitch grid. (b) Photography of the device under test with the microprobes used for the measurements. *Photo courtesy of Xiaoxiong Gu at IBM T. J. Watson Research Center.* (c) 3D full-wave model of the structure. (d) Stackup. The geometrical parameters were defined based on a measurement on a cross-section.

Figure 5.26 compares some of the results obtained by measurement, full-wave simulation, and the proposed models. Reflection at port 3 and near-end crosstalk between port number 3 and port number 2 are shown. Good correlation was obtained up to 20 GHz for the measurement on these two ports. The via crosstalk reaches a high level (> 5 dB) beyond 10 GHz. The near-end crosstalk is plotted up to 40 GHz in Figure 5.27 for other two cases where the port numbers 2-1 and 3-1 were measured. The isolation provided by the ground vias leads to a lower crosstalk in comparison to the simulated case for ports 3-2. The crosstalk S_{31} can become larger than S_{21} at some

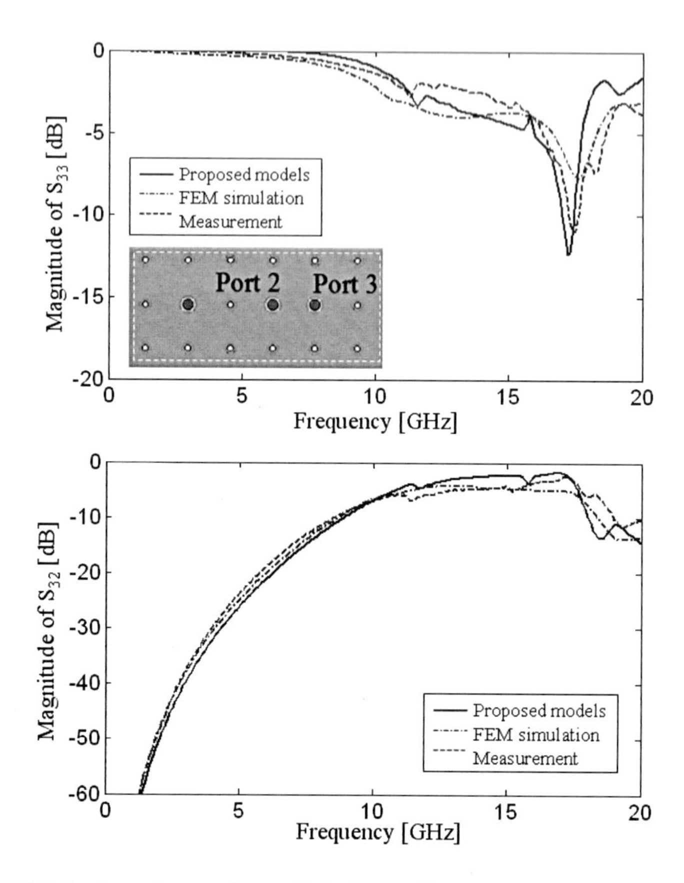

Figure 5.26 Reflection and near-end crosstalk for the 80-mil-pitch via array described in Figure 5.25. Two ports, number 2 and 3, are considered, whereas all the other via ends were left open.

frequency intervals due to the different via environments seen at each port, despite the longer distance between the ports.

The full-wave simulation of such structures took over 12 hours in a multi-core server with interpolation enabled. Due to the model complexity the full-wave simulations were limited up to 20 GHz. In contrast, the solution obtained with the models only required about 40 seconds for 200 frequency points on a single PC (32-bit 3.0 GHz CPU, 4 GB RAM). The same via arrays have been modeled with the foldy-lax scattering method by Gu *et. al* in [136]. Very good agreement with measurements was reported, with a numerical efficiency comparable to the models used in this work.

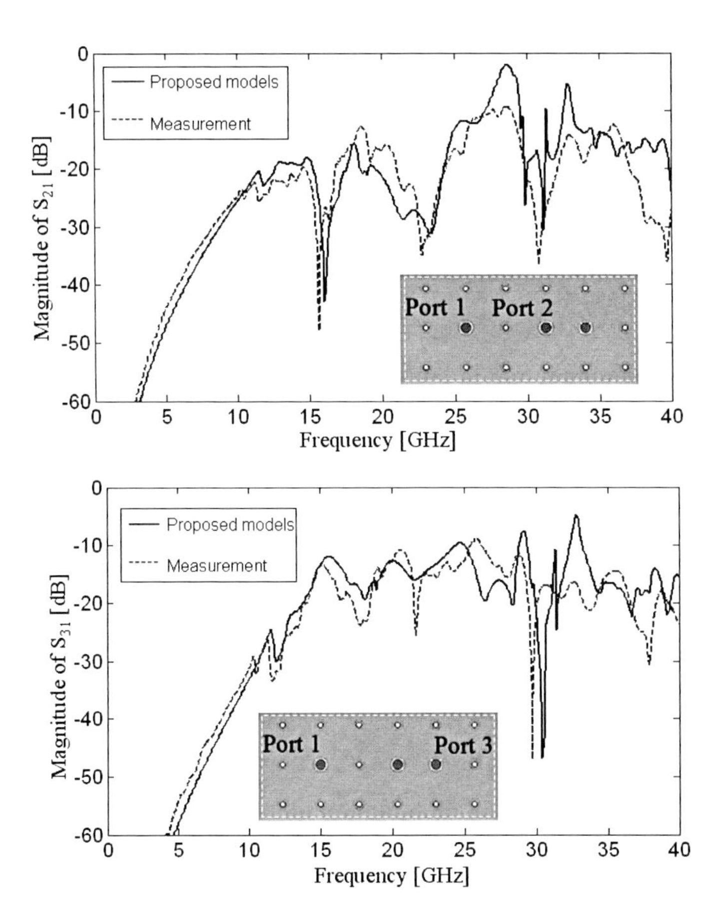

Figure 5.27 Model-to-hardware correlation for near-end crosstalk in a 80-mil-pitch via array. Two different measurements are included, for port numbers 1-2 and 1-3, respectively.

The correlation with denser via arrays has been studied as well. A similar structure to the one described in Figure 5.25, but reducing the via pitch from 80 mil to 40 mil, has been investigated. The resulting agreement is still fair for the measurement related to port number 2 and 3 up to 20 GHz (Figure 5.28).

At higher frequencies, however, the measurements show a different behavior (Figure 5.29). It is possible that scattering effects and near-field coupling among vias may become significant. The results can also be more sensitive to process variations and tolerances. Further studies are required to explain the correlation issues observed for these denser arrays.

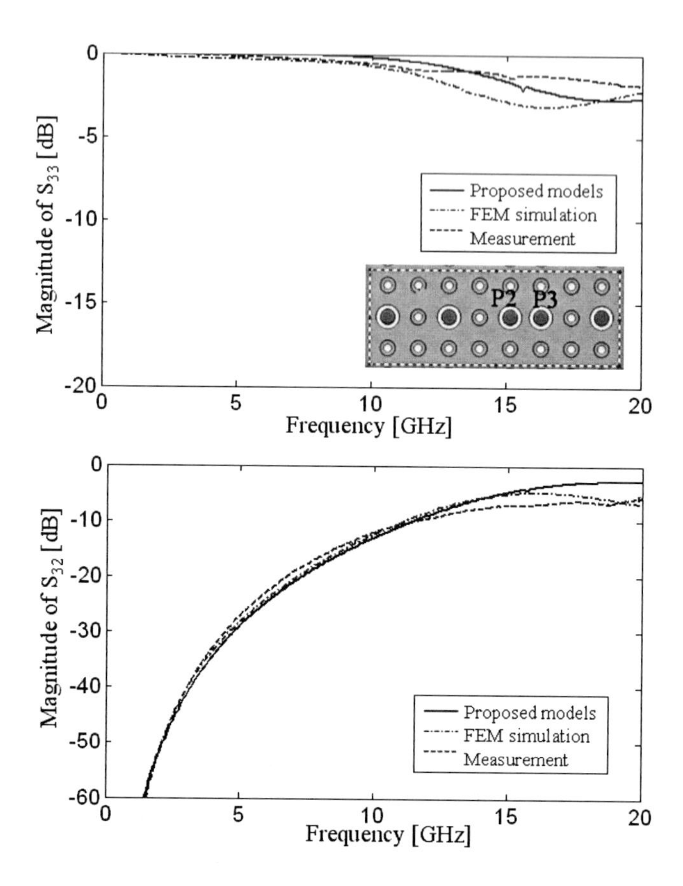

Figure 5.28 Reflection and near-end crosstalk for the 40-mil-pitch via array. Two ports, number 2 and 3, are considered, whereas all the other via ends were left open.

Simulation results are provided in Figure 5.29 for the 40-mil array, assuming PML and PMC boundaries at the board edges. For the PMC case, the plate boundaries were defined at a distance of 1 inch from the outer vias, and the single-summation with 100 iterations was used to compute Z^{pp} with the cavity resonator model. As expected, with absorbing and open boundaries the predicted results are similar up to several GHz due to the isolation provided by ground vias. Nonetheless, it was detected that the models computed with the radial waveguide method (PML case) may start to predict erroneous results with passivity violations. This problem has been observed when many ground vias are placed in close proximity, for multilayer board stacks and at high frequency, typically over 20 GHz. The detailed analysis of this issue and proposed measures to correct it need to be addressed in future work as well.

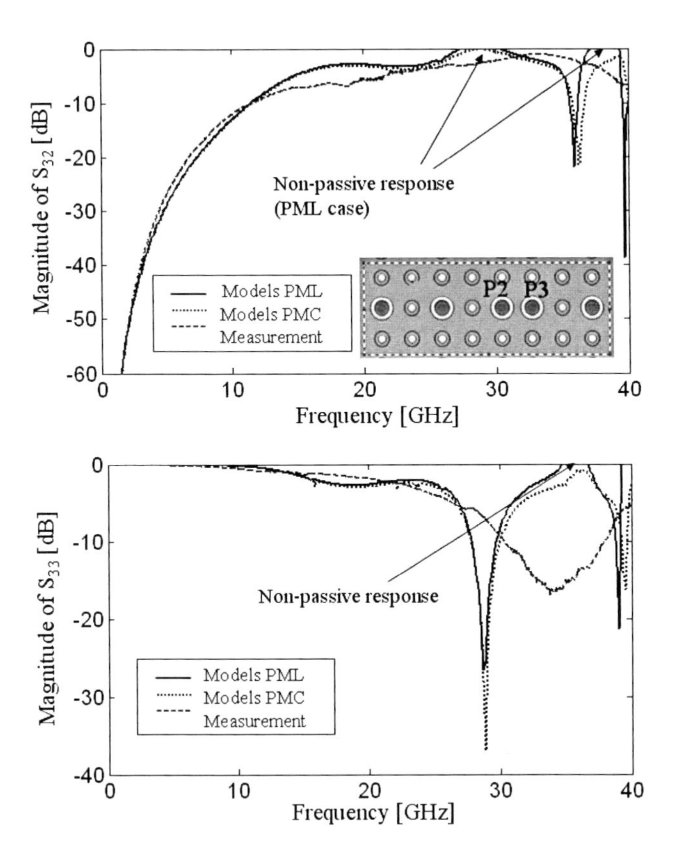

Figure 5.29 Model-to-hardware correlation for near-end crosstalk and reflections in a 40-mil-pitch via array up to 40 GHz. Two ports, number 2 and 3, are considered, whereas all the other via ends were left open. The model results were computed assuming both infinite planes (PML) and PMC boundary conditions at board edges.

5.5. Model Efficiency

One of the major advantages of the proposed models is their high efficiency, which allows the computation of complex structures in a short time when compared to other numerical techniques. The most time consuming part of a simulation is usually the computation of the parallel-plate impedance, followed by the concatenation of model components and cavities. The analytical computation of via capacitances and transmission lines are tasks that can be done quickly. The numerical effort required to get multiconductor transmission line models or to extract capacitances depends on the external solver selected.

The code for simulation of multilayer substrates, which is briefly described in Appendix B, has been developed in Matlab [137]. It is a prototype version that has evolved together with the development of the models and the studies carried out. Note that the current version of the program has not been optimized for speed nor handling of very large configurations, and its further development is recommended.

In order to study the time dependencies, an experiment with via array fields of different sizes has been done. All the simulations have been executed on the same 32-bit PC (3.3-GHz CPU, 4 GB RAM) with a limit of addressable memory of 2 GB per application. The available memory space is however much lower than 2 GB (typically 0.8-1.3 GB) due to other processes running simultaneously and the limited size of contiguous free blocks. With these resources, it is possible to simulate around 1000 via segments and 100 frequency points. This limitation can be overcome by optimizing the code for more efficient memory utilization or by using 64-bit platforms that are able to address more memory.

Figure 5.30 shows the computation time as a function of the number of via elements for a single cavity and 100 frequency points, using the infinite plane formulation for computation of Z^{pp} and fixed values for the via-to-plane capacitances. The simulation times are just approximated since they show a weak dependency on the system load (< 5 %). A program pass refers to the sequence of operations needed, starting from the input description, to compute the overall response of the multilayer structure in the frequency domain. The plot shows two curves that respond to two different manipulations of the frequency points. In the slowest version, each frequency point is computed and stored sequentially (one frequency per program pass). Another alternative is to "vectorize" the frequency points and to compute blocks of frequencies in parallel (several frequencies per program pass). This alternative is more efficient in Matlab, where the vector operations can be executed much faster in comparison to loops [138]. This option allows the computation of the simulation up to ten times faster. Nevertheless, the vectorization of the frequencies requires more memory and is limited due to the available resources. The maximum number of frequencies computed in parallel –on the available platform– decreases from 100 (all frequency points computed on a single pass) to 10 for 400 via segments, to 5 for 600 via segments, and to only one frequency for more than 800 via segments.

The trends in Figure 5.30 also indicate that the increase of the computation time is roughly quadratic with the number of vias. A cavity with 1000 elements requires about 2 hours and 20 minutes, whereas 100 vias can be computed in about 1 minute and 15 seconds.

In Figure 5.31 the computation time as a function of the number of frequencies is studied. A 10x10 via field is used, with one cavity and assuming infinitely large plates.

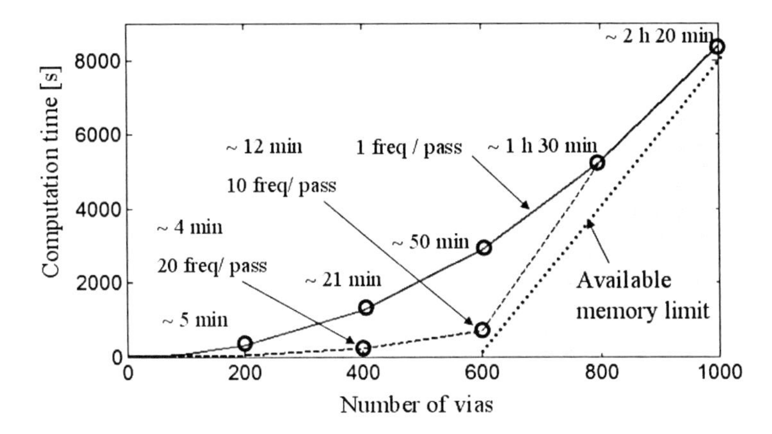

Figure 5.30 Computation time as a function of the number of vias, obtained with the developed code and assuming infinite planes. A single cavity and 100 frequency points were used. The plot shows the case of computing one frequency point and multiple frequencies in parallel per program pass.

The dependency between simulation time and number of frequencies is nearly linear. The manipulation of the frequency points as a vector, computing all the frequencies in one step, improves the computation speed by a factor of 5 to 12, when compared to the case where a single frequency point is computed per pass. For 10 frequencies the problem can be solved in 1.8 seconds with the vector based approach, and in 7.9 seconds when computing a single frequency per pass. One hundred frequencies required about 6.4 and 75 seconds for the same cases, respectively.

The time dependency as a function of the number of cavities is displayed in Figure 5.32. The case corresponds to 100 vias, 100 frequencies and the infinite plate assumption. The simulation time increases about linearly with the number of cavities. More vias lead to longer computation times due to the operations related to the concatenation of partial results, which require the sorting of ports and the application of segmentation techniques. The segmentation requires matrix inversions that start to become time consuming for dense structures.

The algorithm to compute the parallel-plate impedance plays a very important role on determining the overall execution time. One last case, in Figure 5.33, compares the computation time required when the plane impedance is computed with the cavity model single summation with different sum truncation indexes (Eq. (4.10)). The necessary number of iterations, as discussed in Section 4.3.3, increases with the maximum frequency of interest and the plate size. The example corresponds to the same 10x10 via array and 100 frequencies used previously. The computation time grows quadratically with the number of iterations, even though a single sum should reduce

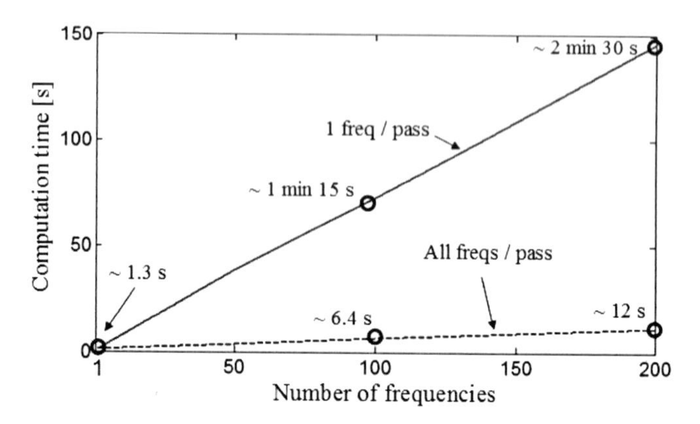

Figure 5.31 Computation time as a function of the number of frequencies with the developed code assuming infinite planes. A 10x10 via array and a single cavity were used. The plot shows the case of computing one frequency point per pass and all the frequencies in one pass.

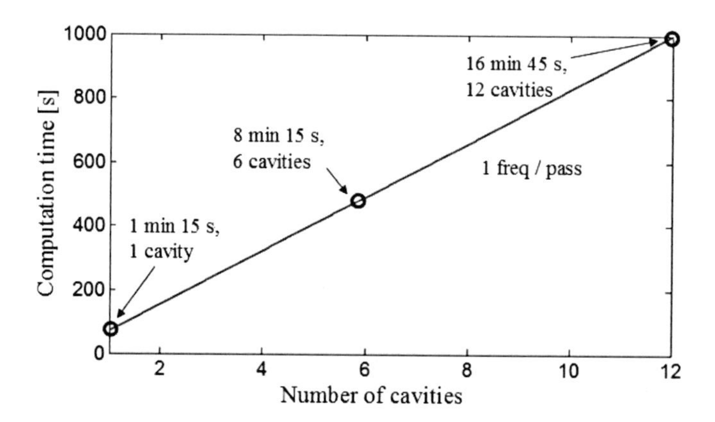

Figure 5.32 Computation time as a function of the number of cavities with the developed code, assuming infinite planes. A 10x10 via array and 100 frequencies, solving one frequency per pass, were used.

the complexity to a linear dependency. This occurs because of the additional program loops that check large argument approximations [109] per iteration, which make the algorithm slower. The efficient coding for computation of Z^{pp} has been addressed in [139]. There, it was shown that by translating the code to a lower level programming language, its efficiency can be improved. With a Fortran [140] version of the routines, a speedup factor of about three could be gained in comparison to the Matlab version.

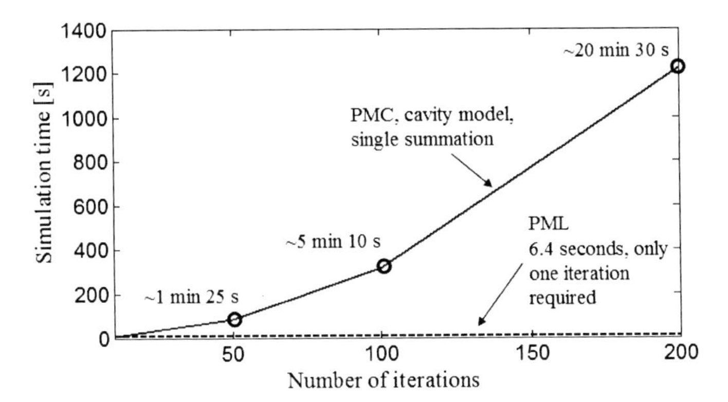

Figure 5.33 Computation time as a function of the number of iterations for the cavity model single summation, for PMC board boundaries, to compute Z^{pp}. The example is a 10x10 via array, a single cavity, and 100 frequencies. As reference the computation time with the infinite plate formulation is provided.

In summary, it has been demonstrated with the discussed examples in previous sections that the proposed models for multilayer substrates show a much higher numerical efficiency when compared to general-purpose numerical methods. The achieved speed-up ranges between two and three orders of magnitude, which can enable fast pre-layout prototyping and optimization of structures with realistic complexity. However, for very large structures with thousands of interconnects the computation times can become prohibitive. Alternatives to further enhance the efficiency of the models should be investigated. There is still room to improve the efficiency of the method by optimizing and writing the code in lower-level programming languages. The parallelization of the code is another option to make a more efficient utilization of the memory and take advantage of multiple processing cores.

Preliminary comparisons have shown that the proposed models offer a comparable efficiency with respect to other hybrid methods. More detailed studies should be carried out to evaluate the advantages and disadvantages of different hybrid approaches, such as the ones reported in [65],[77].

5.6. Model Limitations and Outlook

The results obtained in the preceding case studies –documented in several publications e.g. [11],[14]-[16],[19]– have been satisfactory in terms of accuracy and efficiency. The models have shown to be efficient and flexible enough to handle many structures with a relatively high complexity. Nevertheless, model limitations and potential improvements, most of them already mentioned, have been detected during these analyses. In this section, the main model limitations and recommendations for further development are summarized as follows:

- The analytical formulation to compute the via-to-plane capacitances is available only for vias without pads. In practice, pads are frequently used at via-trace transitions and at via ends. The formulae have shown deficiencies to describe very thin cavities and very short vias. Since the extraction with 3D quasi-static solvers is relatively slow, it is important to investigate alternative approaches that can be used to compute these capacitances more quickly, for arbitrary configurations and in an automated manner. Two dimensional numerical methods that can exploit the cylindrical symmetry of the problem, such as the charge simulation method [141], could be customized for this purpose.
- The via near fields are approximated with simple, uncoupled capacitive elements. At higher frequencies this approximation might become insufficient. For very dense arrays the coupling between vias through evanescent modes, the excitation of higher-order propagating modes in the cavities, and other scattering effects can become significant. These interactions have not been included in the current implementation of the model and they may lead to deviations. In order to address these issues, a more rigorous description of the higher order modes is required, for instance using the analysis of a coaxial-waveguide junction in [97], or by scattering methods [65].
- It is assumed that the ports are small in comparison to the wavelength and therefore their voltage and current distribution can be taken as constant or averaged. Also, it is supposed that the approximations for port shape play a secondary role. These assumptions may start to fail at higher frequencies. In addition, the modeling of irregular structures like vias sharing an antipad has not been addressed yet.
- The models assume that the via excitation fields are isotropic, showing a cylindrical symmetry with no angular dependence. In practice, the vias are sometimes non-uniformly excited, e.g. laterally from a microstrip line or stripline transition, which leads to asymmetrical field configurations that may not be negligible at higher frequencies. It has been observed that for long vias these distortions are less important since they only occur in relatively short sections. For short via transitions the axial field variation can become significant [142].
- The methods to analytically compute the parallel-plate impedance consider the wave propagation between solid and regular-shaped plates with well defined boundary conditions. They do not consider the multiple scattering among ports in very dense arrays. The inter-cavity coupling over board edges and plane perforations such as antipads or slot lines has not been addressed in the current approach. These deficiencies can be partially compensated by alternative

numerical methods and the hybridization of diverse techniques, as discussed in [17], at the cost of a moderate larger numerical burden. Nonetheless, further research is required for the proper handling of these scenarios in multilayer environments.

- The modal decomposition technique assumes that the mode conversion only occurs at via locations. However, it may take place at other locations such as trace discontinuities and traces routed very close to other vias. These cases need to be studied since they are frequently found in board designs.
- The loss mechanisms are described by general formulae and constant material coefficients. More advanced models that can be able to capture the frequency dependencies of material parameters will allow a more accurate solution. The available loss models assume a well developed skin effect and therefore may be inadequate at very low frequencies [104]. The skin effect assumption also neglects the field penetration through planes which is probable at lower frequencies depending on the substrate characteristics.
- The formulation of infinite planes has demonstrated to be useful for efficient computation of complex structures with signal, power and ground vias, within bandwidths of practical interest. However, for very dense arrays with power/ground vias and at high frequencies, passivity violations might occur. The origin of this problem and its correction need to be further investigated.
- The correlation with measurements has shown that the finite impedance of board edges and frequency dependencies of model and material parameters can play a significant role. The importance of process variations and tolerances needs to be further investigated. The accuracy of the calibration methods for measurements and the effect of the signal launches are other topics of interest for the model-to-hardware correlation in the multi-GHz range.

6. Application of the Models to SI, PI, and EMI Analyses

In this chapter the application of the models to diverse case studies related to signal integrity, power integrity, and electromagnetic interference is presented. Special considerations and the extension of the models are addressed together with the examples. Signal integrity studies of differential links, stub vias, and mode conversion are discussed in the first section. Then, the co-simulation of power and signal integrity is covered. The transfer function between signal and power vias, and the effect of surface decoupling capacitors are evaluated. Finally, the extension of the models to co-simulation of signal integrity, power integrity, and radiated emissions is presented. This approach makes use of the contour integral method, which allows the modeling of irregular shapes, and the equivalence principle, to compute radiated emissions at board edges.

The hardware and measurement data presented in this chapter have been provided by the High-Speed I/O Subsystems and Packaging Group at IBM T.J. Watson Research Center [1].

6.1. Application to Signal Integrity Analysis

Three signal integrity studies will be introduced in this section, namely the modeling of differential links between via fields, the prediction of the stub resonance including the analysis of periodic structures made with stub vias, and a study of mode conversion in differential links due to asymmetric ground via configurations.

[1] The hardware and measurements presented in section 6.1.1 were done under IBM contract HR0011-06-C-0074, supported by the Defense Advanced Research Projects Agency (DARPA).

6.1.1. Simulation of Differential Links across BGA Via Arrays

The communication of ICs along boards and backplanes involves the routing of hundreds of I/O nets from the packaged IC. Dense via arrays are used to place the packages and to interconnect the I/O nets at board level. The application of the models to simulate links of realistic complexity has been investigated with the IBM test board in Figure 6.1. Two via arrays designed for ball grid array (BGA) package/sockets are interconnected by multiple differential striplines. The board profile has twelve cavities (although most vias cross only six of them), with 15 cm long traces routed at the S3 layer. With exception of the second power level, the reference planes are assigned to ground. The measurements were carried out with four microprobes placed on top of the defined ports. The probes were previously calibrated on an impedance standard substrate.

In order to restrict the complexity of the problem, the number of modeled vias was reduced to 74 (46 signal, 28 ground), considering only vias in the vicinity of the probed ports, on both the east and west sides. Two links running between the arrays were considered and all the vias not connected to one of the four traces were left open. It was assumed that the vias located further away and the board edges do not play a very important role on determining the response of the measured links. The validity of these simplifications was tested first by computing the plate model by the cavity resonator formulae with open boundaries (PMC) at the board edges (Eq. (4.10)). Then, the results were computed assuming infinite planes according to the radial waveguide method (Eq. (4.11)). The fair agreement between the S-parameters of the cavity resonator model and the radial waveguide method leads to the conclusion that the effect of the boundaries can be neglected in the current example without losing much accuracy. This observation is significant because it implies that, for dense via arrays and test locations placed far away from board edges–which afford an opportunity for return currents to be localized–, the presence of more remote interconnect elements and the size and shape of the board are of minor importance. As mentioned before, the calculation of the parallel-plate impedance assuming infinite planes may noticeably improve the overall efficiency of the simulation, since only one iteration of the algorithm is required per frequency point, in contrast to the cavity method, which requires many iterations to compute the response accurately (see Section 4.3.3).

The transmission line model was obtained with a 2D solver using the cross section information from the stackup. The change in trace widths when entering into the via arrays, the trace coupling between different links, and the effect of trace serpentine bends were neglected. A homogeneous and constant relative permittivity and a constant loss tangent were assumed for the simulations. The conductor loss was considered, but the traces are not centered and thus the modal transformation applied is not exact.

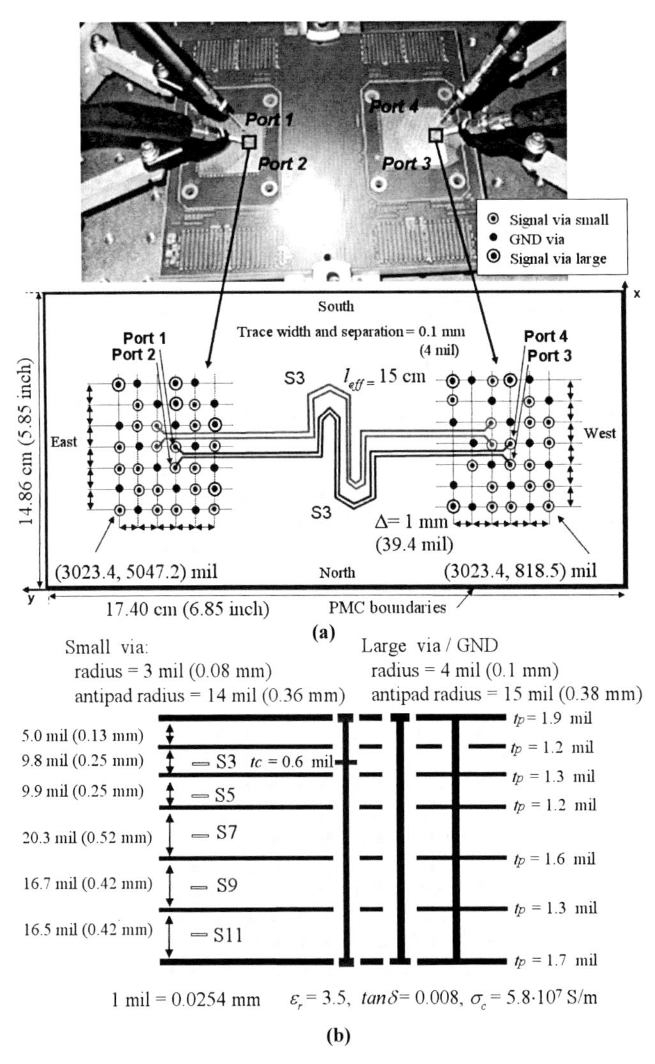

Figure 6.1 Description of the device under test. Two BGA via arrays are connected by differential striplines [19]. (a) Measurement setup and simplified top view diagram of the structure. (b) Simplified board stackup.

Despite the drastic model reductions and simplifications made for the sake of modeling efficiency, the comparison to measurements shows that the models were able to capture the salient features of the hardware response quite well (Figure 6.2). The notch resonance in the transmission parameter (S_{14}), present in both model and

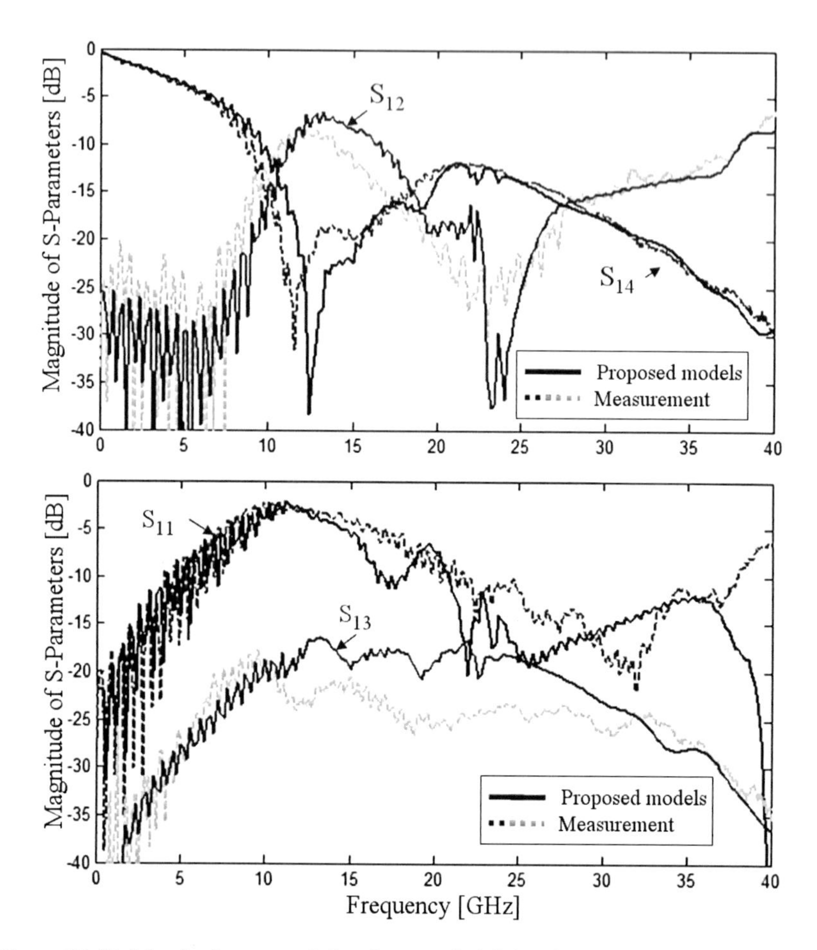

Figure 6.2 Model-to-hardware correlation for a studied link (Figure 6.1). The plots show the magnitude of the single-ended *S*-parameters.

measurement, is a consequence of the via stub length [92]. Possible causes for the model-to-hardware discrepancies are:

- The transmission line models were drastically simplified for the simulation. The board traces present multiple bends and are routed together with many other lines. The change of trace width when the lines are entering into the via arrays, the coupling with other vias, and the skew due to length mismatch were neglected. These simplifications can lead to errors calculating the crosstalk parameters and to underestimate the channel loss, in particular at higher frequencies.

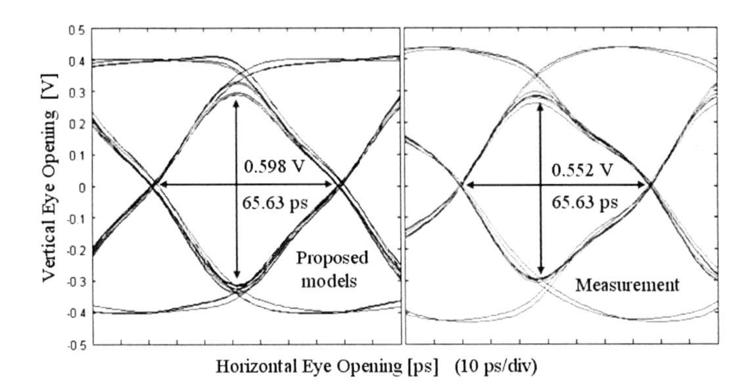

Figure 6.3 Eye diagram for the transmission from port 1 to 4, a data rate of 15 Gb/s and a rise/fall time (t_r 20-80) of 10 ps, obtained with the proposed models and measurement.

- The real board contains many more elements than those being modeled. It was determined, for example, that the resonances present in the simulated results between 20-25 GHz are caused by the vias not connected to a trace. These resonances are dampened when the described structure is extended to consider more adjacent (but still uncoupled) traces.
- Although the microprobes used in the measurement were calibrated, they are likely to introduce additional parasitics that become more important at high frequency. Moreover, the probes on the same link side had to be placed in close proximity, which possibly results in probe-to-probe coupling. These two aspects were not modeled and may impact the correlation, in particular for reflections and near-end crosstalk parameters.
- The models require many parameters that are difficult to calculate with good accuracy over a broad frequency range. The frequency dependencies of the model coefficients and material parameters were also neglected. Process variations and tolerances introduce more uncertainty for parameter estimation. These factors can adversely impact the computation of, for instance, material loss or via capacitances, which may lead to substantial discrepancies in the correlation.
- The via arrays are quite dense and near-field coupling might occur. This interaction has not yet been mapped into the models. In addition, multiple scattering effects between vias and the validity of the cylindrical symmetric field assumption may start to break down for the higher section of the frequency range.

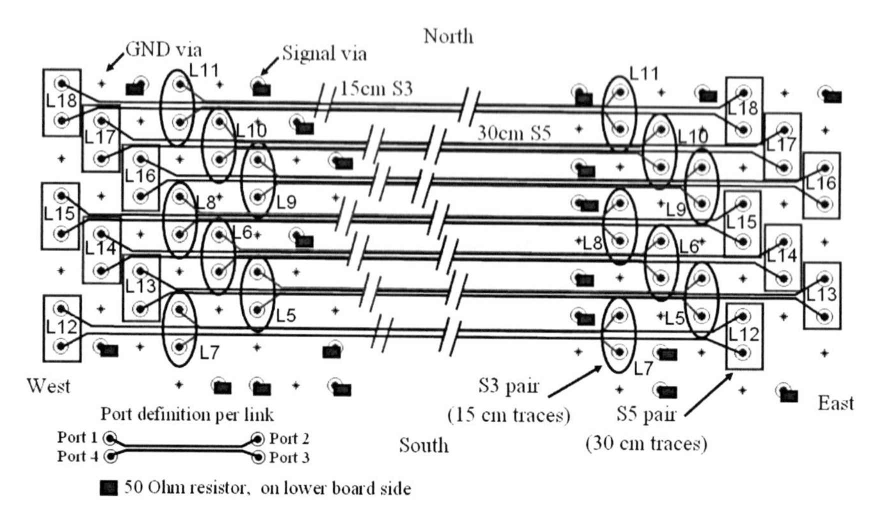

Figure 6.4 Diagram of the extended model for the structure described in Figure 6.1. A total of 119 vias (76 signal and 43 ground vias), 14 differential links (L, seven 15 cm long links routed on the S3 layer, and seven 30 cm long links routed on S5), and 6 cavities were simulated. The stackup is defined in Figure 6.1(b).

The results are satisfactory considering the complexity of the real board structure and the model simplifications. With the simulation results it is still possible to compute the time domain response for data rates up to 15 Gb/s with good accuracy. The shape, vertical and horizontal opening of the eye diagram agree well (within 10 %) for transmission up to that speed, as shown in Figure 6.3.

The computation on a 3.0 GHz 32-bit PC with 4 GB RAM required about 50 seconds, calculating the six cavities independently with 200 frequency points. This value does not include the time required to obtain the 2D transmission line model and to extract the via capacitances. The simulation of even these simplified links is beyond the reach of full-wave simulators running on a single workstation. In contrast, the progress obtained so far with the proposed models has shown that it is feasible to apply them to realistic geometries without incurring onerous computational overhead. Moreover, preliminary comparisons done against commercial hybrid solvers have shown that the proposed method is faster by at least a factor of two.

This case study was extended to consider more differential links, as schematically illustrated in Figure 6.4. Previously, the vias not connected to a stripline were left open. It was determined that this condition introduces some resonances in the S-parameters (see Figure 6.2). Since in the test hardware all signal vias are connected to traces, for the extended model more striplines are considered and the remaining vias were terminated with a resistor on the lower board side to emulate the presence of a

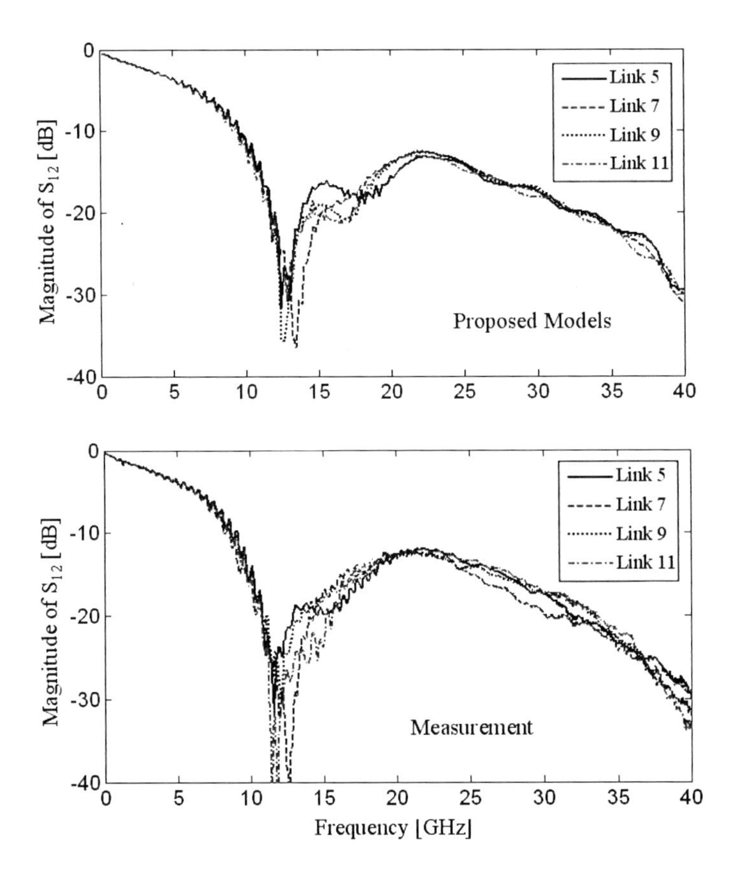

Figure 6.5 Simulated and measured transmission for the extended link configuration in Figure 6.4. The figure shows the single-ended transmission of four 15-cm-long links routed on the S3 level.

trace. The lumped element terminations were incorporated into the models by adding additional entries to the interconnection layers (Section 4.7), which contain the resistance, inductance and/or capacitance of 1-port lumped element.

The link characterized in Figure 6.2 corresponds to the link L5 in the extended structure. The simulation of the new case took about 2 minutes and 47 seconds for 200 frequencies on the same workstation utilized before. Multiple simulations were repeated on different links, since the measurements were done with a 4-port VNA. Figure 6.5 and Figure 6.6 display some selected results (single-ended) where it can be appreciated that the models can capture again the main characteristics of the measured link response. Figure 6.5 shows the transmission for different links routed at the S3 level. The variations are small due to similar via environments for each set of ports, but the same trends for the shift of the stub resonance are observed though. Figure 6.6 provides

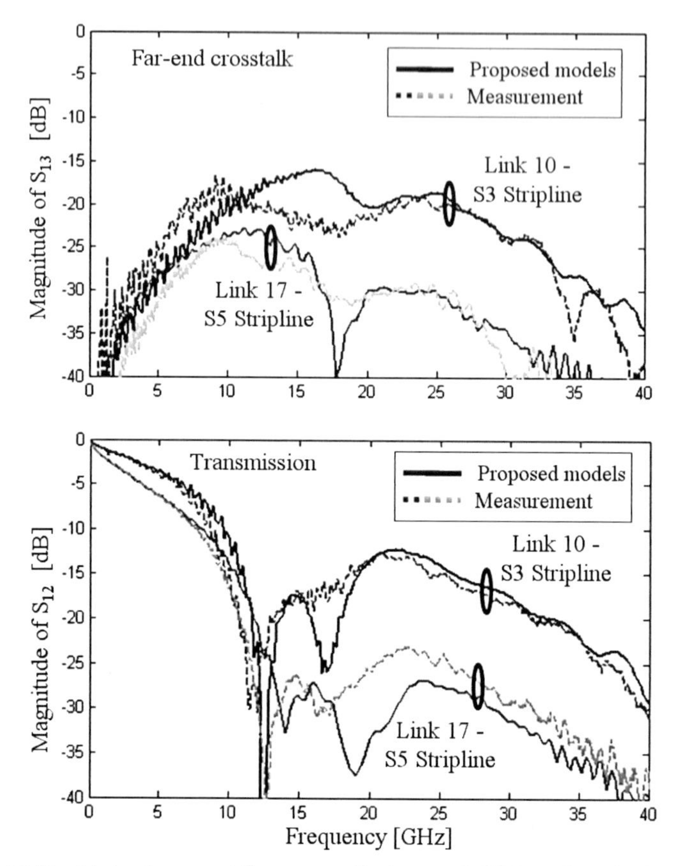

Figure 6.6 Simulated and measured S-parameters for the extended link configuration in Figure 6.4. The plots show the correlation to measurement for the far-end crosstalk and the transmission, for a 15-cm-long link routed at S3 (number 10) and a 30-cm-long link routed at S5 (number 17).

the model-to-hardware correlation for links routed on different signal levels. For a S5 link, the transmission and far-end crosstalk are noticeably lower due to its longer length. Its slightly shorter stub length moves the notch resonance on both the model and measured responses.

The correlation in terms of mixed-mode S-parameters is shown in Figure 6.7 for the link number 5. Differential behavior shadows the effect of some non-idealities, though the agreement is comparable to the one obtained in the single-ended experiments. The models predict a better transmission when compared to the measurement, locating also the stub resonance at a higher frequency. Other studies indicate that mode conversion in the model results is underestimated, mainly due to the idealized 2D transmission line model that does not consider any length mismatch.

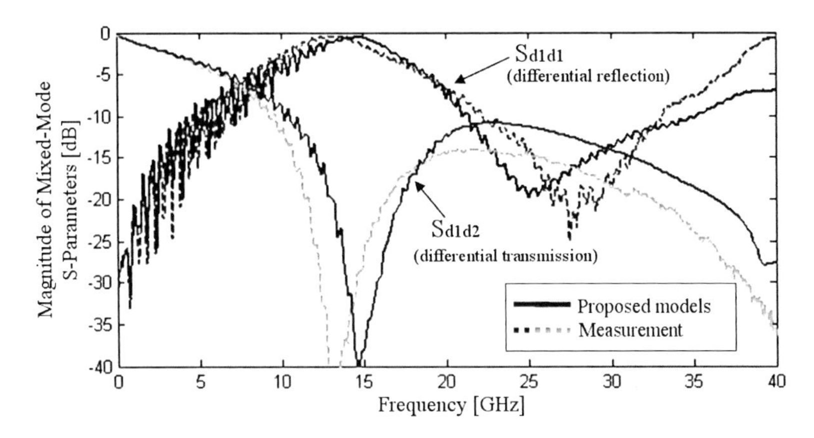

Figure 6.7 Model-to-hardware correlation for the link L5 (Figure 6.4). The plot shows the magnitude of the differential reflection and transmission.

6.1.2. Simulation of Via Stub Resonances

The via stub effect was discussed in Section 3.6, where it was shown that the stub length introduces unwanted notches in the transmission along links in multilayer substrates. The proposed models have been used to simulate stub via configurations with two different signal-to-ground via ratios, as depicted in Figure 6.8. As before, the measurement was done with a VNA and calibrated microprobes. Since the goal was to measure the stub resonance of a single via per link, one probe is positioned on top of the via and the other at the trace end by means of a recessed probe launch (RPL). A RPL is made by milling the board layers down to the trace and placing the probe directly on the trace. The launch has ground vias and pads in close proximity in order to provide a good return path to the probes [26],[128]. In the models, the RPL was approximated by a short buried via in close proximity to two return vias.

The separation between all adjacent vias was equal to 40 mil (Figure 6.8(b)). Absorbing boundaries at the board edges were used, though the real hardware is a large panel with several other test structures nearby. The transmission line models and via capacitances were computed externally with a 3D solver.

The transmissions of the first links for cases 1 and 2 are provided in Figure 6.9. The first case shows a sharper resonance located at about 12.5 GHz. There is good

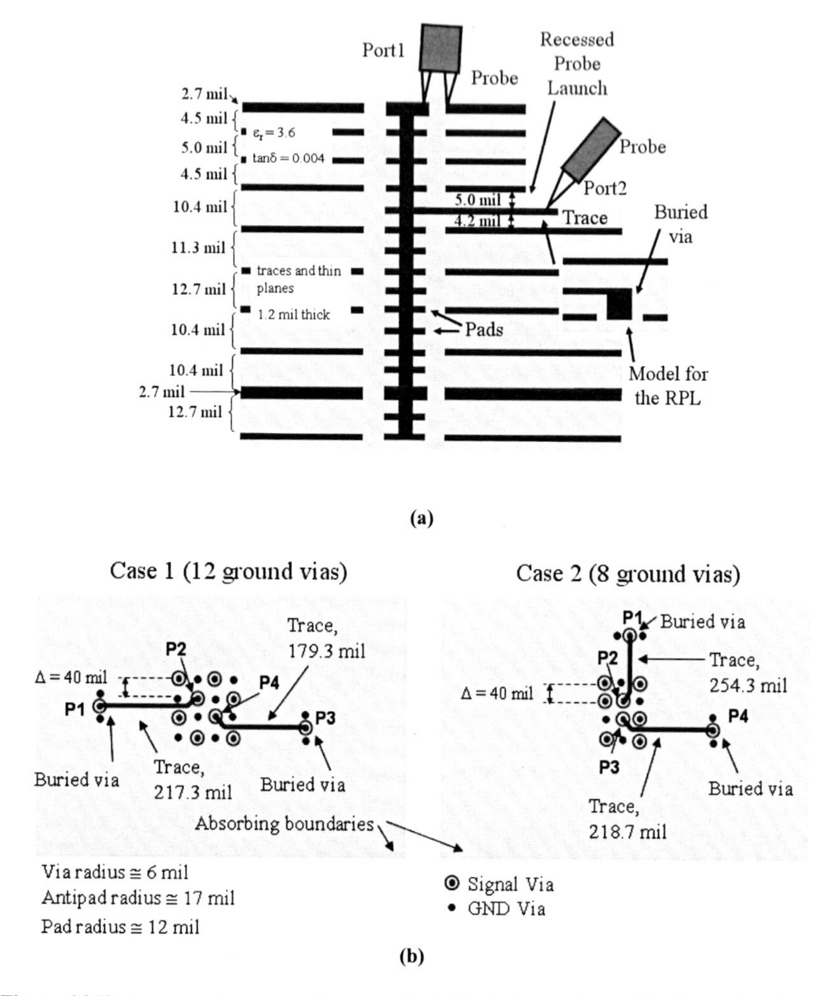

Figure 6.8 Test structure to evaluate the stub effect. (a) Stackup and port definition. All reference planes were assigned as ground. The recessed probe launch used in the physical structures was approximated by a short buried via in the models. (b) Top view of the via configurations. The pitch between adjacent vias is 40 mil.

agreement between the measurement and models for the transmission up to 20 GHz. The effect of via pads on the notch is also shown. Without the pads the stub resonance is shifted towards higher frequencies. The pads increase the via capacitance, which tends to make the wave propagation on the via slower and therefore the equivalent

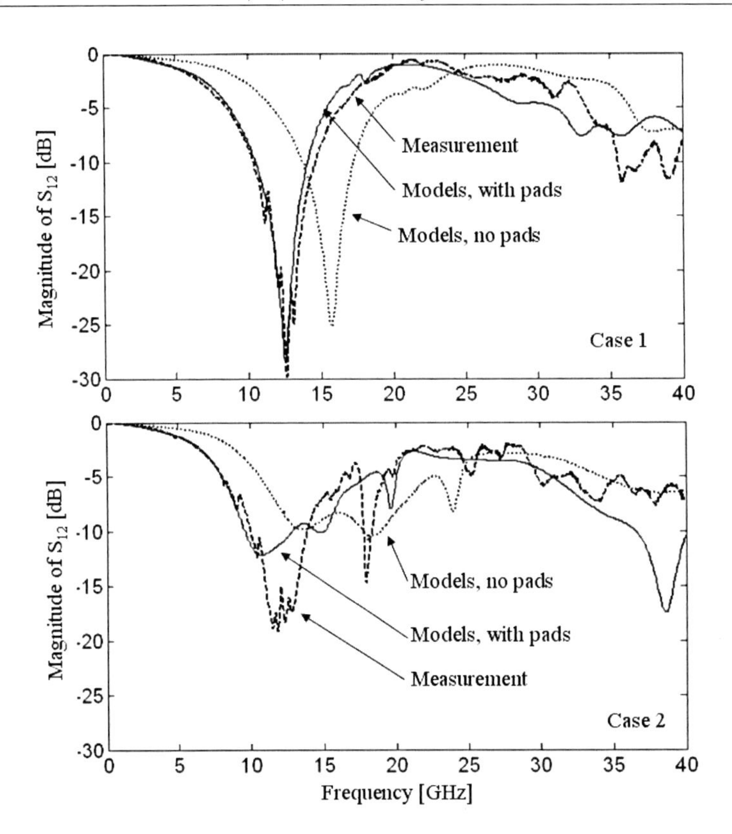

Figure 6.9 Correlation to measurement for cases 1 and 2 in Figure 6.8. The plots include the effect of via pads on the transmission parameters.

stub electrical length longer. For the second case the stub resonance tends to be shifted to lower frequencies due to a smaller number of return vias. In the models the notch is dampened which is attributed to the absorbing boundaries used in the simulation. With fewer ground vias the waves can travel more freely inside the cavities. As seen in Section 3.1, the infinite plate case results in larger loss, which is reflected as a dampening of the notch resonance in this configuration. For the test vehicle, less ground vias result in more interference from other neighboring structures which may impact the results and the correlation. Note that all these observations made in this study were consistent with the trends discussed in Section 3.6.

Although the stub effect is detrimental for high-speed digital signals traveling on multilayer substrates, it may be exploited for some microwave applications. For instance, a stub configuration may serve as a narrow-band coupler [143], or periodic stub via structures can be used to design filters [144].

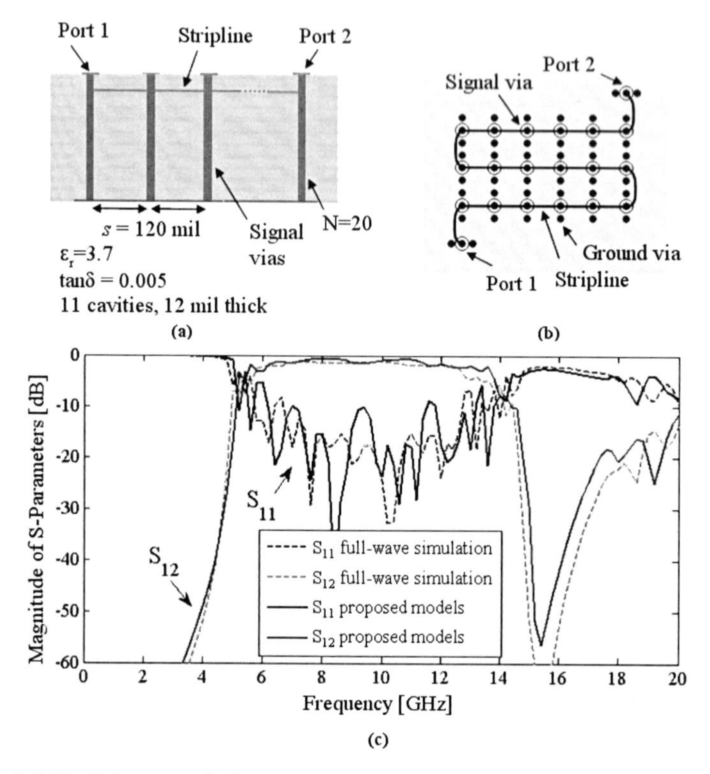

Figure 6.10 Simulation of a periodic structure with stub vias. (a) Stackup diagram. The via radius and antipad radius were 5 mil and 15 mil, respectively. (b) Top view diagram of the structure. (c) Simulated *S*-parameters obtained with the proposed models and a FEM full-wave simulation.

Figure 6.10 describes an example of a periodic structure with a chain of 20 stub through-hole vias connected by traces at the second cavity. The via pitch is 120 mil and the stub ends are terminated with a short circuit. This configuration provides a passband between approximately 4 and 15 GHz. Good correlation is observed for the full-wave results and the ones obtained with the models up to 20 GHz.

This configuration has the peculiarity that one via is connected to two traces. This case is of limited interest for links in multilayer substrates, but it might be useful for other scenarios. This special via-to-stripline transition can be easily handled by realizing that the traces contacting the same via resemble a parallel connection. According to the notation of Figure 6.11, the transmission line model (Y^{tl}) defined at the three via-stripline transitions assuming ideal-ground planes is required (see Section

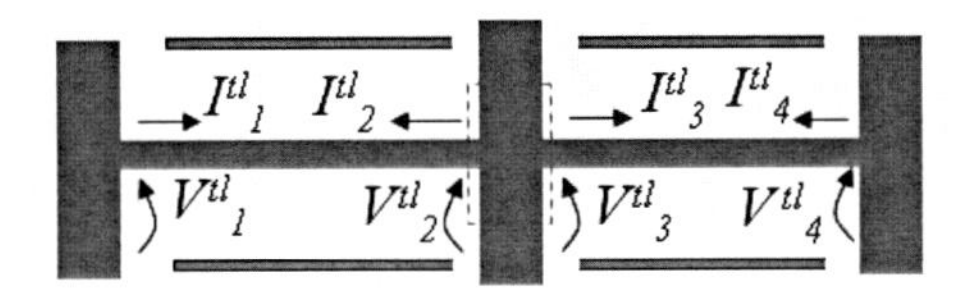

Figure 6.11 Simplified description of the case of a via connected to multiple traces. The model considers the parallel connection of the two traces.

4.6). For each trace the terminal voltages and currents are defined according to Figure 6.11, and the admittance matrices for each trace have the form

$$\begin{pmatrix} I_1^{tl} \\ I_2^{tl} \end{pmatrix} = \begin{pmatrix} Y_{11}^{tl} & Y_{12}^{tl} \\ Y_{21}^{tl} & Y_{22}^{tl} \end{pmatrix} \cdot \begin{pmatrix} V_1^{tl} \\ V_2^{tl} \end{pmatrix}, \tag{6.1}$$

$$\begin{pmatrix} I_3^{tl} \\ I_4^{tl} \end{pmatrix} = \begin{pmatrix} Y_{33}^{tl} & Y_{34}^{tl} \\ Y_{43}^{tl} & Y_{44}^{tl} \end{pmatrix} \cdot \begin{pmatrix} V_3^{tl} \\ V_4^{tl} \end{pmatrix}. \tag{6.2}$$

Assuming that the voltage drops at the trace terminals number 2 and 3 are identical, for the center via-stripline transition the following equations apply

$$I_{eq}^{tl} = I_2^{tl} + I_3^{tl}, \tag{6.3}$$

$$V_{eq}^{tl} = V_2^{tl} = V_3^{tl}. \tag{6.4}$$

Terminal numbers 2 and 3 can then be combined to form a single one,. From Eqs. (6.1) -(6.2) and (6.3)-(6.4), and as Y-parameters, this can be expressed as

$$I_{eq}^{tl} = Y_{21}^{tl} \cdot V_1 + (Y_{22}^{tl} + Y_{33}^{tl}) \cdot V_{eq}^{tl} + Y_{34}^{tl} \cdot V_4. \tag{6.5}$$

The reduced transmission line model is the 3-times-3 admittance matrix

$$\overline{\overline{Y}}_{eq}^{tl} = \begin{pmatrix} Y_{11}^{tl} & Y_{12}^{tl} & 0 \\ Y_{21}^{tl} & Y_{22}^{tl} + Y_{33}^{tl} & Y_{34}^{tl} \\ 0 & Y_{43}^{tl} & Y_{44}^{tl} \end{pmatrix}. \tag{6.6}$$

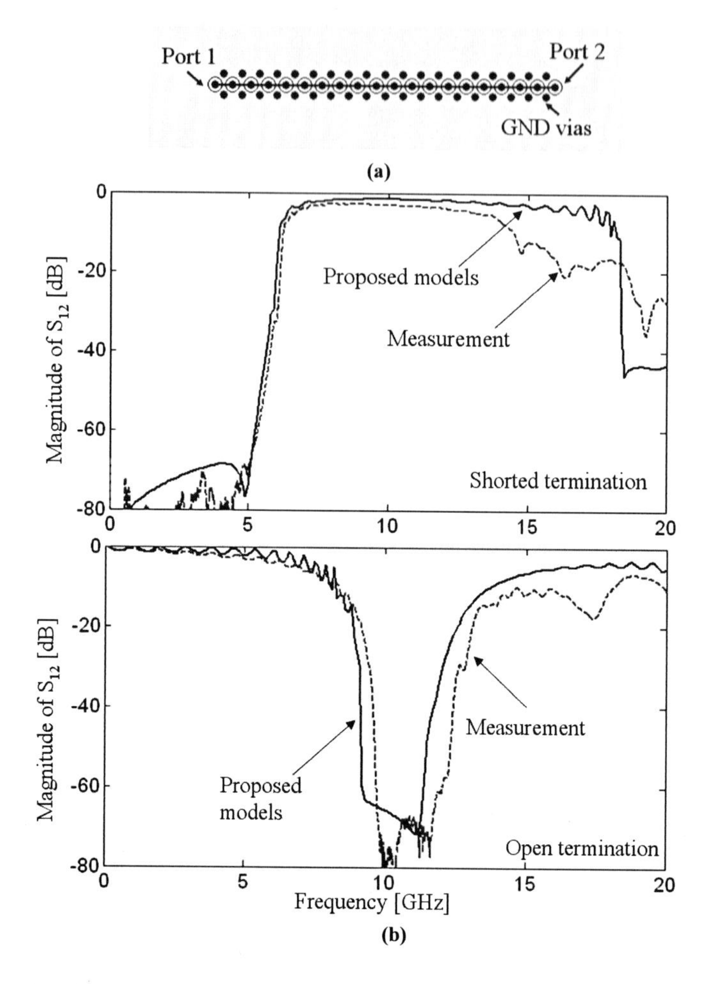

Figure 6.12 Model-to-hardware correlation for an array of periodic stub vias. (a) Diagram of the structure. It contains 20 signal vias connected by traces at the second cavity, with 100-mil separation. The stackup is defined in Figure 5.25 and the via and antipad radius are 5 and 15 mil, respectively. (b) Transfer function when the stub vias are terminated with a short circuit and an open circuit.

In terms of Y-parameters only the addition of the input parameters connecting the same via is required. This is applicable to more traces contacting the same or different vias, assuming that the traces remain uncoupled to each other and the distortion in the near field due to the traces is negligible. The combined transmission line model should be coupled to the parallel-plate model by modal decomposition. Different k factors should be defined for each transition if the traces are not aligned (e.g. if two signal levels exist inside the same cavity).

Model-to-hardware correlation studies using IBM test vehicles have been carried out for similar structures to the one described in Figure 6.10. The stackup of the boards corresponds to the one discussed in the via arrays in Section 5.4 (Figure 5.25). The configuration is a row of 20 signal vias, with a pitch of 100 mil, and connected by traces at the second cavity (S1 level). Two cases are considered with short and open via stub terminations, which provide a pass-band or reject-band (notch) filter characteristic, respectively. Figure 6.12 indicates that both the models and the measurement predict the filter band and show fair agreement up to 10 GHz. The differences at higher frequencies are attributed to model simplifications and the influence of non-modeled adjacent structures. The ground vias do not provide enough isolation in this configuration and the distances to other structures and board edges are not negligible. Further investigation on the utilization of the models for microwave applications is suggested as future work.

6.1.3. Simulation of Mode Conversion in Differential Links

Differential signaling is widely used in today's high-speed digital system designs for its relative good immunity to noise and crosstalk. However, the differential mode (DM) signal can be converted into common-mode (CM) signal by various mechanisms such as skew and rise/fall time mismatch from IC driver, signal trace length mismatch, and asymmetric configurations [145]. Mode conversion can degrade the system signal integrity. Moreover, even small amounts of common-mode signals can have significant impact on the EMI performance [146].

The models are utilized in this section to analyze the DM to CM conversion in via-to-via differential links with different ground via configurations. This study will address the impact of asymmetric ground vias in the vicinity of a link and how they can increase mode conversion as much as trace length mismatch. The possibility to mitigate mode conversion by compensating the trace length mismatch with asymmetric ground vias and vice versa is explored as well.

The reference link structure is illustrated in Figure 6.13. Two through-hole via pairs are connected by a 2-inch-long coupled trace. The trace width and gap between traces are set to 4 mils. The stackup consists of 12 cavities, each of them of 12 mil thickness.

The dielectric is assumed to be homogeneous ($\varepsilon_r = 3.8$, $\tan\delta = 0.03$) and the metallic conductor regions are modeled with a conductivity of $\sigma = 5.8 \cdot 10^7$ S/m. The selected via pitch is 40 mil. The via radius is equal to 5 mil and the antipad radius is 15 mil. It is assumed that the board edges are located far away and the associated reflections are negligible. The differential insertion loss of the link was simulated for traces routed in

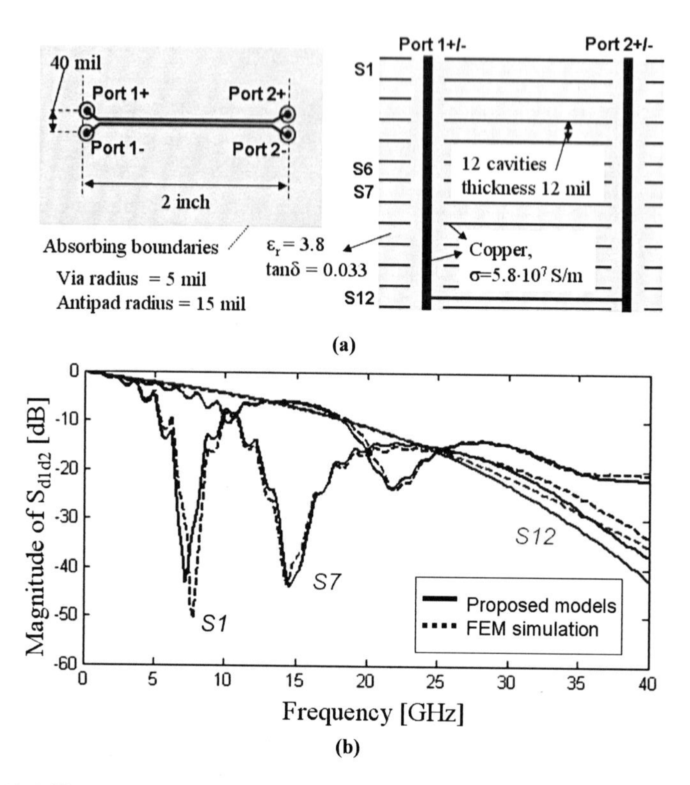

Figure 6.13 Differential link structure under investigation. (a) Link diagrams. (b) Simulated differential transmission placing the coupled traces at different signal levels, obtained by the proposed models and full-wave simulation.

different cavities –S1, S7 and S12, identified on the stackup– and therefore possessing different stub lengths (Figure 6.13(b)). Good agreement can be observed between the results obtained by the semi-analytical models and full-wave simulations.

The simulation of this link configuration took over 10 hours with the FEM method running on a 3-GHz CPU with 4-GB RAM. The solution using the proposed models can be computed in less than 15 seconds with the same resources, which allows the study of many scenarios in a very short time.

The relation between the DM/CM conversion and the asymmetric placement of ground vias with respect to the signal vias was studied. Figure 6.14 depicts an example adding two ground vias and a set of selected results. The agreement between the results

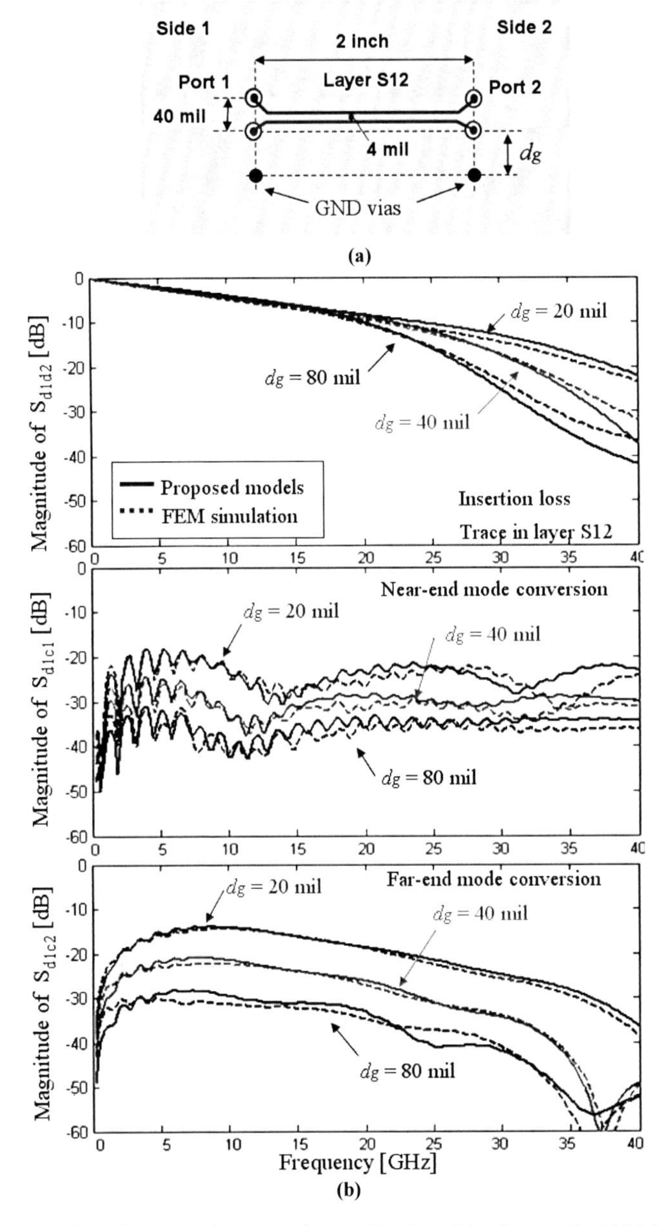

Figure 6.14 Effect of asymmetric ground vias as a function of the distance *dg*. (a) Link configuration (trace at S12, stackup of Figure 6.13). (b) Simulated insertion loss and mode conversion.

Side 1 Side 2
2 inch
Layer S12

Asymmetric ground via configuration	S_{d1c1} @ 5 GHz (dB)	S_{d1c2} @ 5 GHz (dB)
1 GND on side 2	-34.4	-26.6
1 GND on side 1 / 1 GND on side 2	-24.3	-21.8
2 GND on side 1 / 2 GND on side 2	-22.5	-19.6
3 GND on side 1 / 3 GND on side 2	-22.1	-18.6

Figure 6.15 Differential to common-mode conversion for different asymmetric ground via configurations. The distance to the nearest signal via *dg* was 40 mil in all cases, with the trace routed at the S12 level.

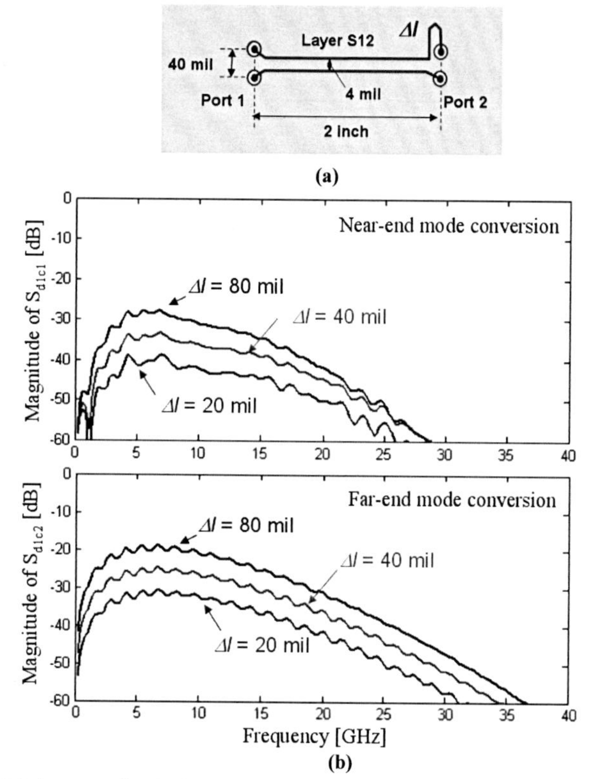

Figure 6.16 Mode conversion in the differential link due to trace length mismatch: (a) link diagram, (b) far and near end mode conversion as a function of the length mismatch Δl.

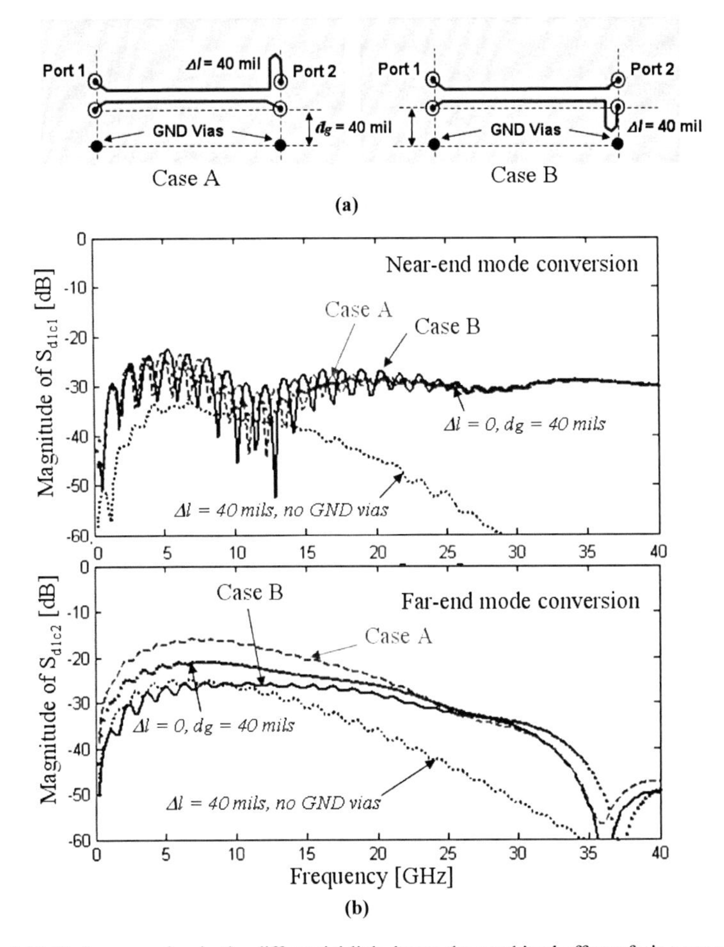

Figure 6.17 Mode conversion in the differential link due to the combined effect of via asymmetries and trace length mismatch: (a) studied cases, (b) far- and near-end mode conversion.

computed with the models and the full-wave analysis is also good for mode conversion in their mixed-mode S-parameter form.

The amount of mode conversion was computed as a function of the distance from the center of the ground via to the center of the nearest signal via dg. Figure 6.14 plots the magnitudes of the differential insertion loss Sd1d2, the near-end mode conversion Sd1c1, and the far-end mode conversion Sd1c2.

It can be observed that Sd1d2 is only weakly influenced by the variation of the distance *dg* at frequencies below 20 GHz. Beyond 20 GHz, the proximity of ground vias to the signal vias tends to improve the differential signal transmission. However, the behavior of the mode conversion parameters Sd1c1 and Sd1c2 strongly depends on the distance *dg* in almost all the studied bandwidth. The closer the ground via is to the signal via, the greater the mode conversions. This is due to the fact that the structure looks more asymmetric as the ground via is placed closer to the signal via. Therefore via asymmetry can have a great impact on the mode conversion, while having little effect on differential-to-differential signal transmission.

The mode conversion can be larger than -20 dB, e.g. for the case $dg = 20$ mils from 3 to 20 GHz in the far end. Nevertheless, the influence of ground vias on mode conversion decreases with distance. When the ground via is located at a distance larger than two pitches away (2 x 40 mil) the mode conversion is below -30 dB and its effect could be neglected. Several other scenarios have been studied. Figure 6.15 shows the DC/CM conversion for different asymmetrical cases with ground vias placed near to the lower signal via. The larger the asymmetry introduced by ground vias the higher the mode conversion. Note that all the symmetric simulations done predict values below -50 dB.

Trace length mismatch between the two traces of a differential link constitutes another cause of mode conversion. For comparison purposes, the DM/CM conversion for this case was investigated as well. The configuration is shown in Figure 6.16, where Δl represents the length mismatch at the far-end port 2 and is modeled as an extra uncoupled transmission line segment. As expected, the mode conversion is higher for a longer mismatch Δl, and it decays rapidly as the frequency increases beyond 10 GHz due to channel loss. The mode conversion levels –from -40 dB up to -15 dB for the studied configurations– are similar when compared to the cases of asymmetric ground vias.

Both the via distribution and trace length mismatch can contribute to DM/CM conversion in a similar proportion. The interaction between the two mechanisms depends on the link configuration. The combined effect on mode conversion was studied for the two cases shown in Figure 6.17. Case A considers two ground vias asymmetrically placed on the lower link side, and a trace length mismatch introduced on the top link path. Case B presents a similar structure, but the trace length difference is located on the lower link path. Figure 6.17(b) compares the near-end and far-end mode conversion parameters of cases A and B, as well as the cases of only ground vias, and only length mismatch. It is observed that near-end mode conversion *Sd1c1* is at a similar level for both cases A and B due to the asymmetric ground via placing. However, far-end mode conversion Sd1c2 is up to 10 dB higher for case A than for case B at frequencies below 20 GHz.

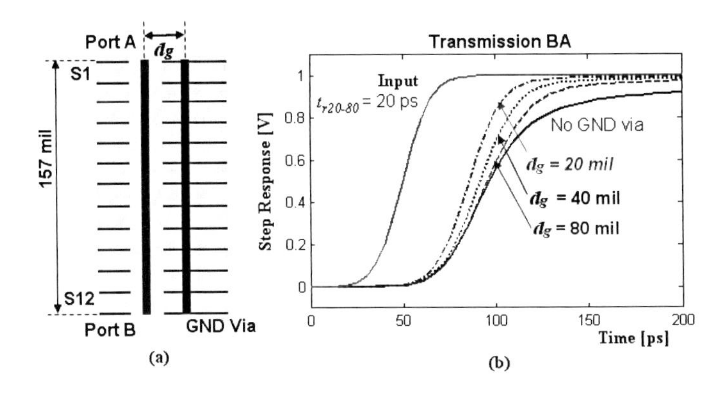

Figure 6.18 Effect of a ground via as a function of the separation between the signal and ground via. (a) Through-hole via case with a ground via. (b) Simulated step response for different via separation *dg*.

Heuristically, this behavior can be explained by the fact that the ground vias tend to reduce the equivalent via inductance and, hence, to increase the propagation velocity of the signal along the via. The closer the ground via to the signal via, the shorter the equivalent electrical length of the via appears. In case A, the ground vias influence mainly the lower link side, which tends to become electrically shorter than the top path. Since an extra trace segment is introduced for the top link path, the imbalance is even stronger and this leads to higher mode conversion. In contrast, for case B the electrical length reduction introduced by the ground vias is partially compensated by trace length mismatch on the same side. This observation is also consistent with the results displayed in Figure 6.14, where mode conversion increases as the distance to the ground via is reduced.

Figure 6.18 shows this effect in time domain. The step response of the signal traveling along a via is plotted as a function of the distance *dg*. The results show that the signal tends to propagate faster and the rise time becomes shorter as the ground via is placed closer to the signal via. This effect appears also in cases A and B as a skew increment or reduction between the signals traveling on each link side, respectively.

The examples discussed before indicate that when asymmetries cannot be avoided, the via positions and trace lengths can be carefully selected to compensate opposite effects and to minimize mode conversion.

6.2. Application to Combined Signal and Power Integrity Analysis

The power distribution network interconnects defined by the power/ground vias and planes, and the signal nets formed by the signal vias and traces become tightly coupled at high frequencies. Signals traveling over vias penetrate the power planes and excite parasitic modes supported by parallel plates. Similarly, power noise coupled into the PDN can propagate and interact with signal nets (Figure 6.19).

In this section, it will be demonstrated that the models are general enough to indistinctively handle power/ground and signal nets and that they are therefore suitable for co-simulation of both power and signal integrity domains in a comprehensive and efficient manner. Figure 6.20 depicts the schematic representation of the models for a single cavity bounded by a top ground plane and a bottom power plane. Two vias connected by a trace are represented, where the blocks denoted as Tmd stand for the modal transformation matrices. For power/ground vias the connectivity of the vias to the planes is incorporated. When a via is connected to a power plane, the via capacitance in the model is replaced with a short circuit. This simple consideration allows the modeling of structures with arbitrary power/ground via and plane configurations. Additional elements, such as decoupling capacitors (decaps), can also be included by adding their network representation to the corresponding entry in the matrices that contain the information of the via-to-plane capacitances and plane connectivity (see Section 4.5). The decaps have been modeled as lumped RLC networks, as indicated in Figure 6.20(b).

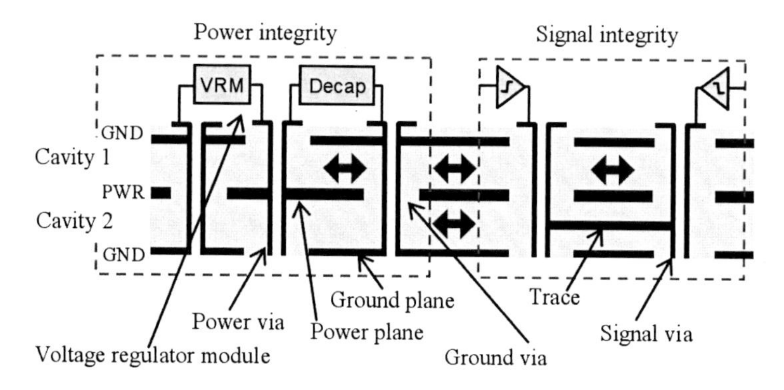

Figure 6.19 Conceptual identification of the power and signal integrity domains in multilayer substrates.

6.2.1. Simulation of the Interaction between Signal and Power Vias

The structure analyzed in this section is described in Figure 6.21. Two differential links, each of them formed by four through-hole vias and a differential stripline, are routed on a six-cavity substrate with mixed power and ground planes. The fourth level is defined as a power plane, whereas the other six planes are assigned to be ground. Additional ground and power vias are placed in the vicinity of the signal vias, for a total of 28 elements. Ports were defined on top of the 8 signal vias (ports 1 to 8) and two additional ports were included as observation points on top of power vias (ports 9 and 10). The simulation of the example was performed using both the presented semi-analytical models and a FEM solver. Some of the results obtained with the two techniques are displayed in Figure 6.22 and Figure 6.23. For instance, in Figure 6.22 both the model and the full-wave analysis predict a worse differential transmission for the second link (ports 3 to 6). This is attributed to the position of the traces at a higher level with respect to the other link. The longer via stubs shift the notch in the transmission to lower frequencies.

The configuration selected can also be used to study the interaction between signal and power nets and consequently to explore the two main mechanisms responsible for noise injection in multilayer structures: the currents being injected or supplied by the PDN to other devices such as ICs and the crosstalk between signals routed on the substrate [147].

Figure 6.23(a) shows the single-ended crosstalk between a power via (port 9) and a signal via (port 2), and between two signal vias (ports 3 and 2). The crosstalk of the power and the signal nets tends to be larger in comparison to the crosstalk of two signal nets for frequencies below 14 GHz. The port transfer functions are influenced by several factors such as position or isolation due to ground vias [148]. The difference observed in the current example can be explained by the fact that the signal via is located farther from port 2 and the trace guides a portion of the signal to other elements, which reduce the crosstalk. Nevertheless, both coupling mechanisms between vias are significant enough to degrade signals in the GHz range. The results have also been translated into the time domain using an inverse Fourier transform. The impulse responses for the signal at port 2, due to a 1-V 100-ps Gaussian pulse applied at the PDN port 9 and the signal port 3, are plotted in Figure 6.23(b). The induced voltage amplitudes, up to 2 % and 0.5 % of the input pulse, may not be negligible, in particular if it is considered that in a real scenario the total noise is a result of the simultaneous interaction of many nets.

Another possibility to analyze the results is to display them in terms of generalized mixed mode S-parameters [149] (see Appendix A.8), in order to evaluate the impact of differential or single-ended signaling. Figure 6.24 shows the crosstalk and the impulse

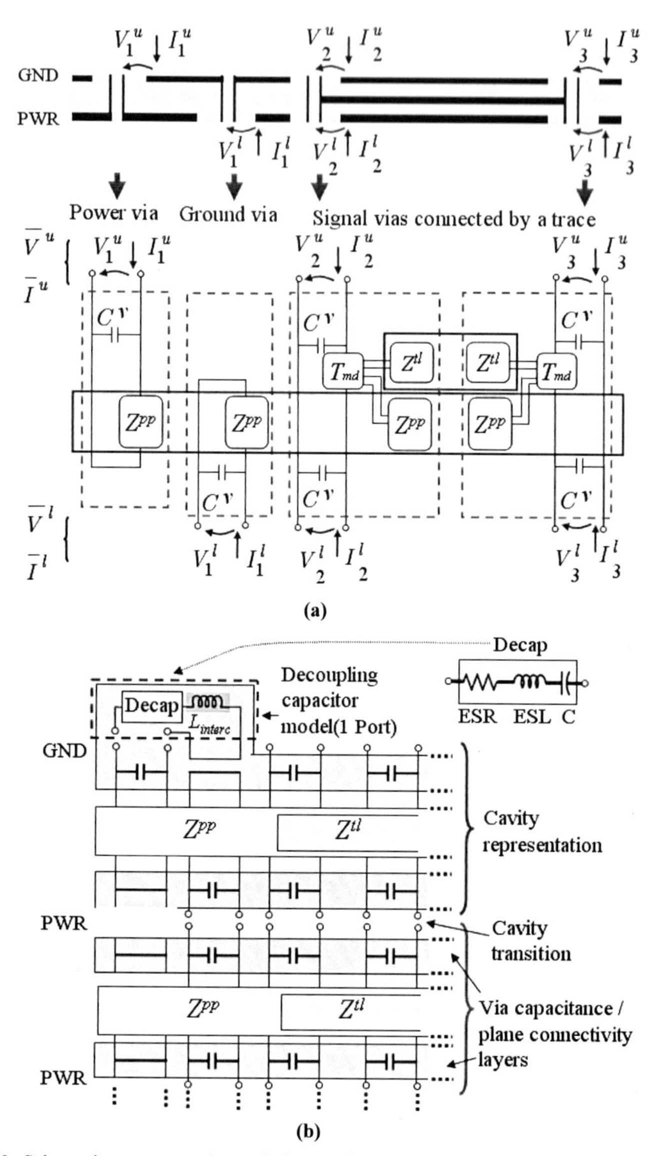

Figure 6.20 Schematic representation of the models to include PDN elements. (a) Network representation of different via types in a single ground-power cavity. (b) Model concatenation for multilayer substrates including decoupling capacitors. These capacitors are modeled with a lumped RLC network that includes the capacitance value, the equivalent series resistance (ESR), the equivalent series inductance (ESL), and the inductance associated with the interconnects required for their placement (L_{interc}).

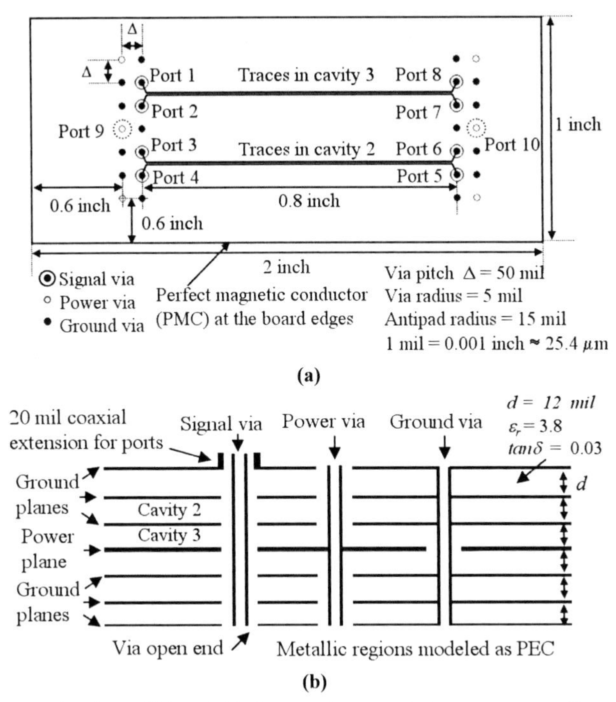

Figure 6.21 Example of two differential links routed on a multilayer substrate with mixed reference planes (one power plane). (a) Top view. (b) Cross section.

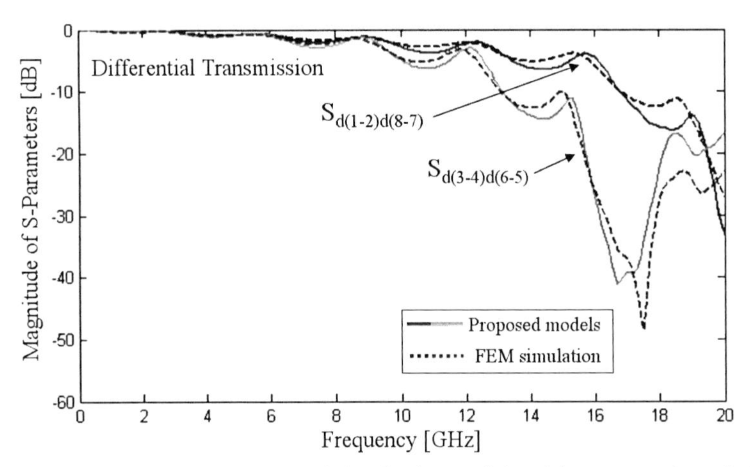

Figure 6.22 Simulated differential transmission for the two links of the structure shown in Figure 6.21, obtained with the proposed models and full-wave simulation.

response from the power port 9 (*se: single-ended*) to the differential pair formed by ports 1 and 2 (identified as *c: common-mode* and *d: differential-mode*). The plots indicate that the differential-mode signaling can noticeably improve the immunity to noise, at least considering ideal differential sources.

The simulation of the test structure took about 21 seconds for 200 frequency points in Matlab, running on a 3.0-GHz CPU with 4 GB of RAM. The computation of the parallel-plane impedance was done once using the cavity model with single-sum implementation [105] and 100 iterations per impedance term. The capacitances were calculated analytically using the formulation in [118], adding a 20 fF capacitance to compensate the fringing fields at open via ends. In contrast, the elapsed time for the full-wave analysis exceeded 40 hours using 8 adaptive passes for a target convergence of 2% and 200 frequencies, running on the same computer. This represents a speedup factor of about 6800. With the models, however, if every cavity is different and a separate computation of Z^{pp} is required per layer, the simulation time can become longer. For the studied case, the analysis considering an independent computation per cavity took about 94 seconds. Nevertheless, this still represents an improvement of the computation speed of over three orders of magnitude.

6.2.2. Simulation of Structures with Decoupling Capacitors

As an extension of the example discussed in the previous section, the power and ground via configurations are modified in order to quantify the impact of ground vias (Figure 6.25). The simulation of the new configurations took approximately 9, 12, 40, and 180 seconds for case I (16 vias), case II (20 vias), case IV (42 vias) and case V (90 vias), respectively. The port definition, geometrical and material parameters correspond to the ones defined for the previous example in Figure 6.21, which is now case III.

The crosstalk for different power/ground configurations is plotted in Figure 6.25(b). For cases I and II, the predicted crosstalk is larger due to the small number of ground vias. In contrast, for cases III, IV and V the results show a very similar crosstalk level between the power and the signal ports. A similar trend can be observed for the crosstalk between the signal nets in cases IV and V. By taking a closer look at the configurations, it can be stated that larger crosstalk variations occur when nearby vias are altered. This experiment shows that adding many power/ground vias will not always lead to a noticeable improvement on the response and that it is important to carefully select the position and number of such vias.

Another alternative to reduce the coupling between vias through the power planes involves the utilization of surface decoupling capacitors. These capacitances are incorporated into the simulation method as additional lumped elements connected to

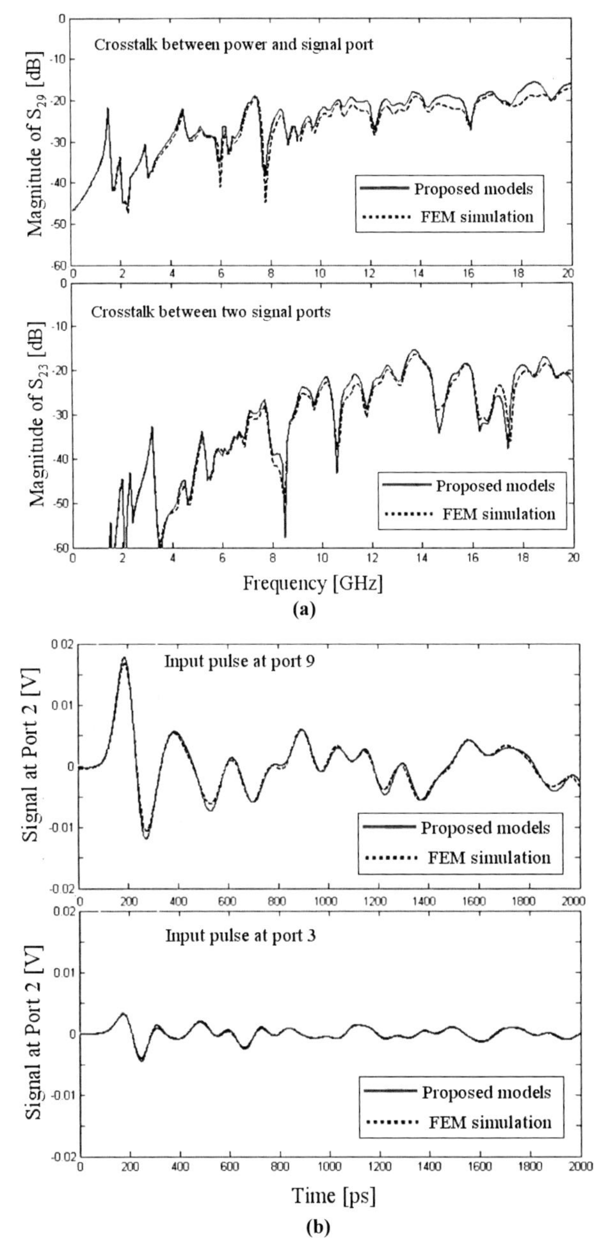

Figure 6.23 Single-ended crosstalk between a power and a signal via, and two signal vias, according to the port definition of Figure 6.21. The graphs show the results obtained with a full-wave analysis and the semi-analytical models.

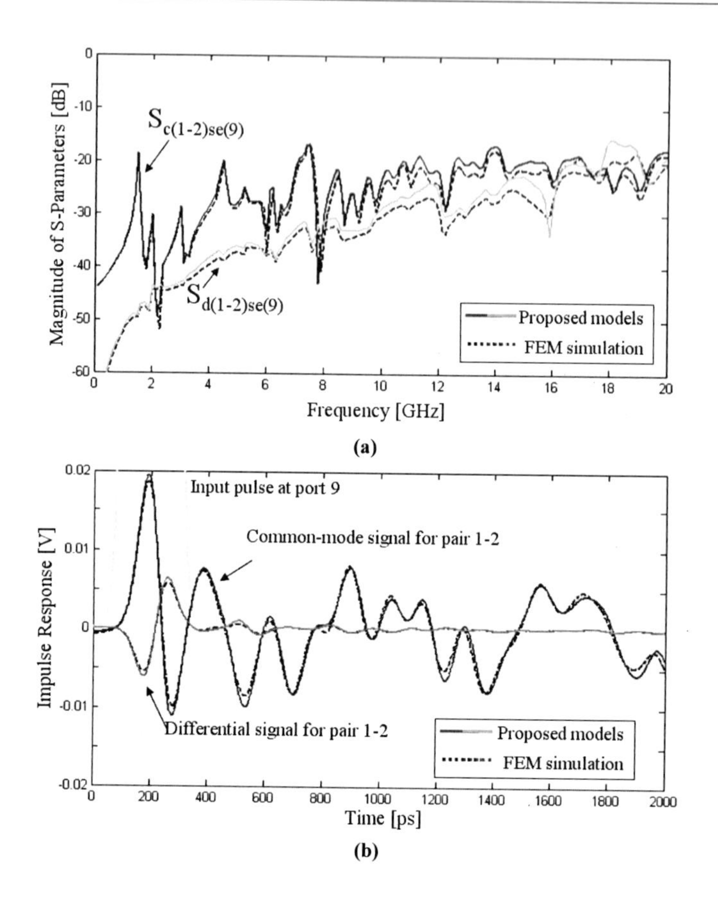

Figure 6.24 Differential and common-mode interaction between a signal and a power net. (a) Differential-mode and common-mode crosstalk. (b) Impulse response at pair 1-2 for a 1-V 100-ps full-width-half-maximum Gaussian pulse applied at port 9. The graphs show the results obtained with a full-wave analysis and the semi-analytical models.

the via ports (Figure 6.21). For the current set of examples, decoupling capacitors were connected on top of all power vias, with exception of the observation ports 9 and 10. These elements were modeled as an RLC network to represent the capacitor equivalent series resistance (ESR), the equivalent series inductance (ESL), and the inductance of external interconnects required to place the capacitor on top of the vias (L_{interc}) [48]-[49]. All the capacitors have the same values: ESR = 100 mΩ, ESL+$L_{interc.}$ = 2 nH, C = 10 nF.

The decoupling effect is more often visualized in Z-parameters. Note that the impedance reduction due to decaps can be related to a crosstalk reduction between signal and power vias as well. Figure 6.26 shows the transfer impedance for two power

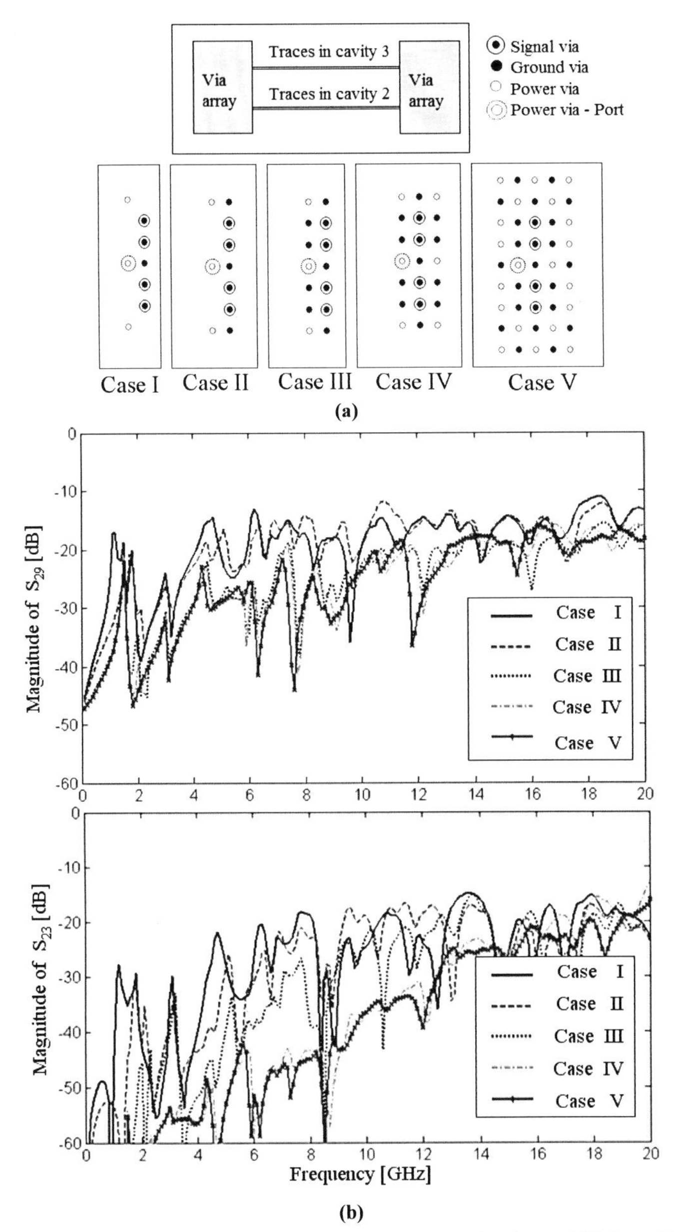

Figure 6.25 Alternative configurations for the reference case in Figure 6.21 (case III). (a) Link definition and via array configurations (same array used at both link sides). (b) Crosstalk between a power and a signal port, and two signal ports, for different power via configurations.

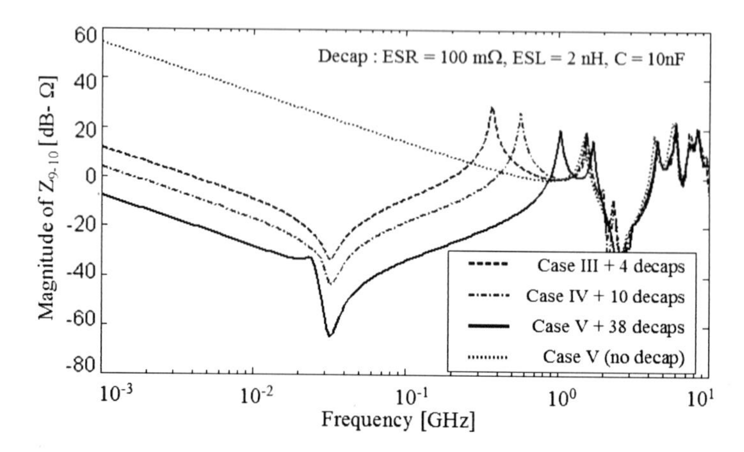

Figure 6.26 Effect of decoupling capacitors on the transfer impedance between two power vias, for different configurations of Figure 6.25 and a different number of decoupling capacitors.

vias with 4 decaps in case III, 10 in case IV, and 38 in case V. For the studied cases, the effect of the capacitors is observable mainly at frequencies below 1 GHz. The cavity resonant frequencies are located at relatively high frequencies because of the board size. A stronger decoupling is also difficult to achieve due to the small cavity thickness, which translates into a low impedance that can become comparable to the parasitics of decaps. The results indicate that more capacitors help to increase the equivalent capacitance and to reduce the equivalent inductance, increasing the maximum frequency where the decoupling may be effective. The first parallel resonance arises when decaps are placed due to the introduced parasitic inductances.

In Figure 6.27 the effect of the parasitics associated with decoupling capacitors is shown for the case V. Different positions of the power plane are simulated, with ideal and non-ideal decaps. The graph shows that pure capacitances connected to the second power plane can provide the largest reduction of the plane impedance at higher frequencies. The decoupling effect is reduced when the power plane is changed to a deeper layer. When the ESR, ESL and L_{interc} are added, the effect of the capacitors tends to be further diminished. The distance to the plane and the parasitic elements associated with surface capacitors make effective decoupling in multilayer environments difficult [49]. Moreover, the punctual nature of surface decoupling capacitors makes them unable to cancel propagating waves once they are spread out from their sources.

Additional studies that have addressed the further application of the models for co-simulation of power and signal integrity, and including model-to-hardware correlation, have been carried out and presented in [11].

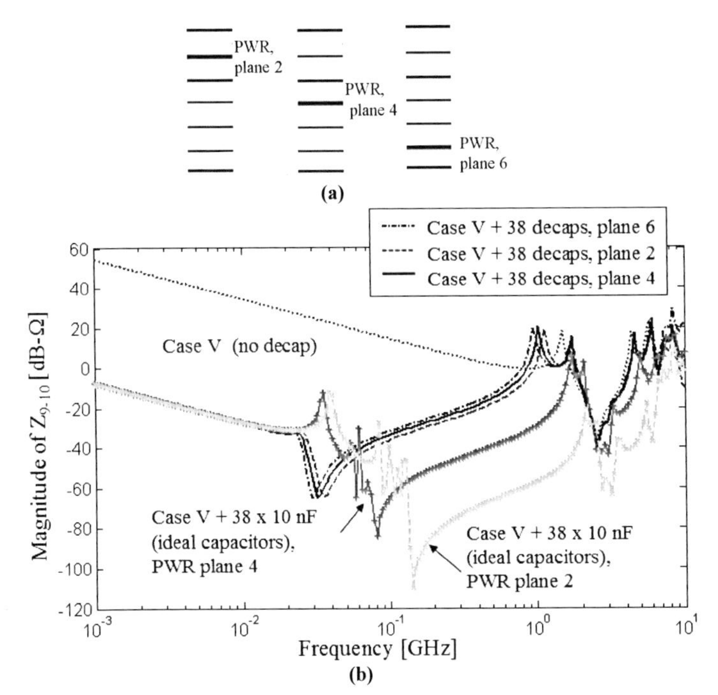

Figure 6.27 Effect of parasitic elements associated with decoupling capacitors. (a) Power plane assignment for the simulations, which are variations of the configuration in Figure 6.25. (b) Transfer impedance between two power vias for different power plane positions and capacitor values. The decaps were modeled as an RLC networks with ESR = 100 mΩ, ESL+ $L_{interc.}$ = 2 nH and C = 10 nF. The two impedance lines with markers consider ideal capacitors without ESR, ESL, or L_{interc}.

6.3. Application to Combined Signal and Power Integrity, and Radiated Emissions Analysis

Signal integrity, power integrity, and electromagnetic compatibility of digital systems are three domains that are closely related. For instance, signal transitions among different layers can excite cavity modes of power planes and cause power-ground bounce [150]. Noise can be coupled to other vias or cause radiated emissions through the plane edges. As the complexity of the interconnects increases and the rise/fall time reduces, their interactions become more prominent. Disregarding them may cause costly and less optimal noise suppression in a late stage of the design flow. Therefore, the trend is to unify the analyses of signal integrity (SI), power integrity (PI), and radiated emissions (RE).

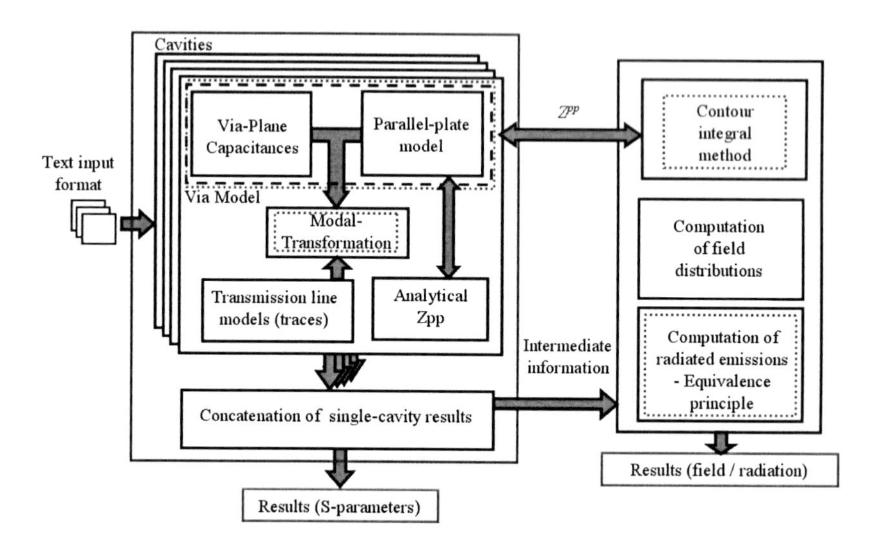

Figure 6.28 Block diagram of the framework for co-simulation of signal integrity, power integrity and radiated emissions. The four main components are the via and trace model, modal decomposition, the contour integral method, and the equivalence principle.

This section presents a method for fast and comprehensive simulation of multilayer substrates including metrics of these three domains by combining the merits of the via model, modal decomposition, the contour integral method, as well as the equivalence principle. The via model in combination with modal decomposition and segmentation offer an efficient technique for signal integrity analysis of multilayer substrates. The contour integral method (CIM) can be used to calculate the voltage distribution between arbitrarily shaped power planes (Appendix A.9). Far-field radiation can be obtained by applying the field equivalence principle (Appendix A.10). The four techniques are applied to analyze SI, PI, and RE in a fast, concurrent, and holistic manner. Figure 6.28 shows the extended simulation approach, including the CIM and the equivalence principle. This work was done in collaboration with another project at TUHH, where the CIM related code was developed [17].

6.3.1. Extension of the Method for Analysis of Multilayer Printed Circuit Boards

The previous proposed models and the simulation method have been applied to describe multilayer structures for SI and PI analyses [12],[19]. They have been focused on obtaining the end-to-end port response and the electromagnetic behavior at intermediate layers was not of concern. The matrices describing each cavity (Eq. (4.38)) are transformed and concatenated to obtain the overall S-parameter response. The method was then extended to compute the noise distributions between power/ground planes and the far-field radiation for PI and EMC applications. For these purposes, it is necessary to obtain the currents flowing on all the via segments of each intermediate layer and the field distributions on the cavity boundaries. With the CIM only the board boundaries need to be discretized, which in principle simplifies the meshing and potentially results in better numerical efficiency in comparison to other 2D numerical methods.

The modeling of individual cavities, including both via and boundary ports has been considered. Then, the partial results are cascaded together using the segmentation technique [78] to obtain the multilayer response. Figure 6.29 shows the cross section of a 5-layer 3-cavity PCB example and the equivalent network and port definitions. The block diagram of the most important elements and the coupling mechanisms for the second cavity is included in Figure 6.30. It is assumed that adjacent cavities are only connected through the via transitions. The capacitances C^v model the via near field. The interaction between vias and the power planes is described by the parallel-plate impedance matrix Z^{pp}. Striplines connecting the vias are modeled with the impedance Z^{tl}. The transfer impedance Z^{pq} represents the noise propagation to the cavity boundary, which induces the radiated emissions. The impedance matrix obtained by CIM corresponds to a cavity including both via and boundary ports, defined by the superscripts p and q, respectively (Appendix A.9)

$$\begin{bmatrix} \overline{V}^q \\ \overline{V}^p \end{bmatrix} = \begin{bmatrix} \overline{\overline{Z}}^{qq} & \overline{\overline{Z}}^{qp} \\ \overline{\overline{Z}}^{pq} & \overline{\overline{Z}}^{pp} \end{bmatrix} \cdot \begin{bmatrix} \overline{I}^q \\ \overline{I}^p \end{bmatrix}. \tag{6.7}$$

As described in Section 4.1, the via ports are expanded to account for the top and bottom layer connections. This can be carried out in various parameter forms. It was found more straightforward to address the via port expansion using an h-parameter expression, defined as

$$\begin{bmatrix} \overline{\overline{h}}^{qq} & \overline{\overline{h}}^{qp} \\ \overline{\overline{h}}^{pq} & \overline{\overline{h}}^{pp} \end{bmatrix} \cdot \begin{bmatrix} \overline{I}^q \\ \overline{V}^p \end{bmatrix} = \begin{bmatrix} \overline{V}^q \\ \overline{I}^p \end{bmatrix}, \tag{6.8}$$

where the h-matrix can be obtained from the impedance matrix in Eq. (6.7) by the transformation [124],[151]

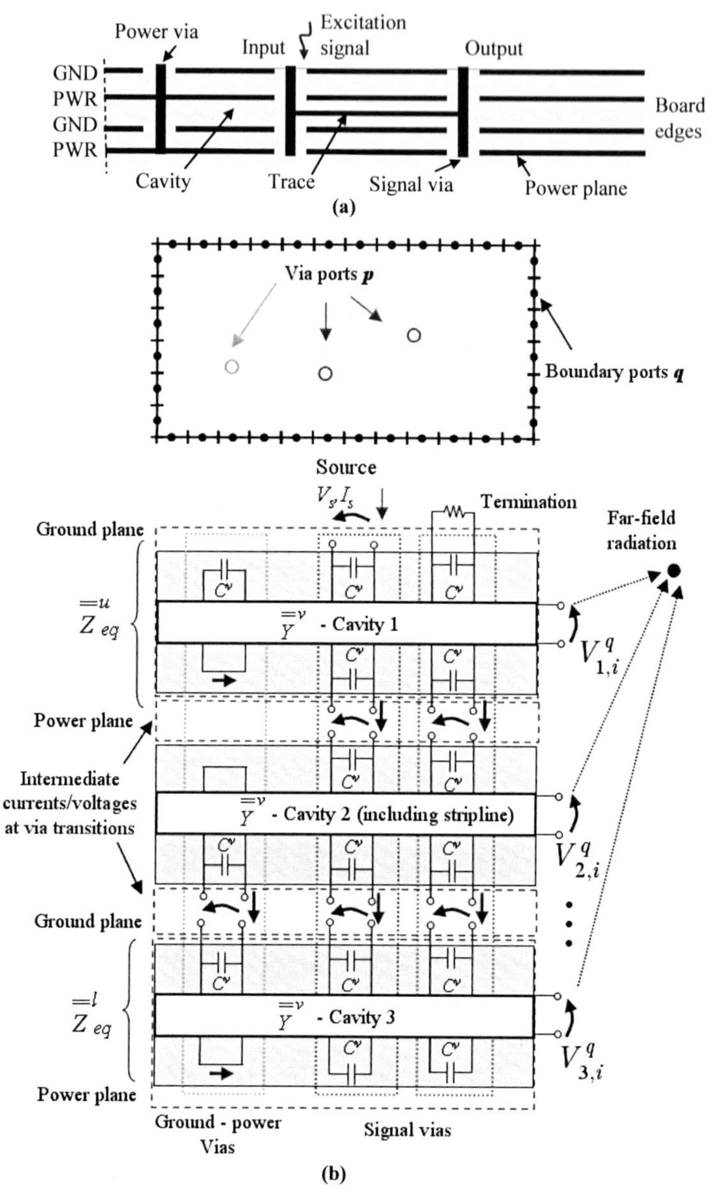

Figure 6.29 Description of the combination process for a multilayer case. (a) Example of a board cross-section with three cavities. (b) Network-level representation of the example, including the port definition for via and boundary ports. The via and trace connections are handled inside the cavity blocks.

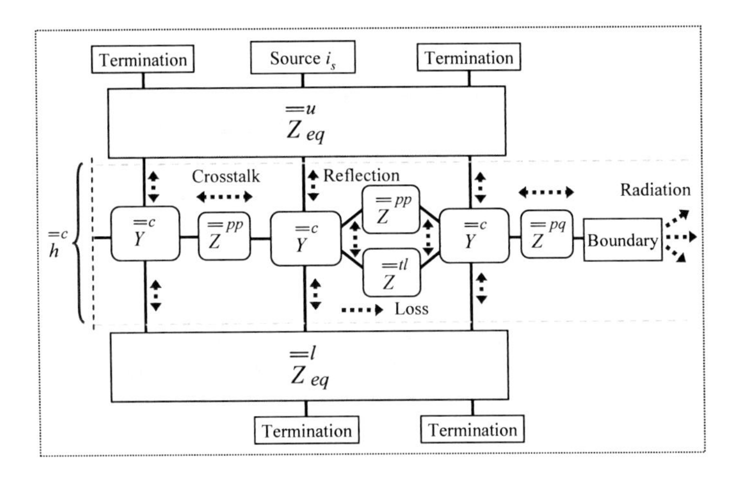

Figure 6.30 Equivalent block diagram for the second cavity in the example of Figure 6.29, taking into account the contribution of the other cavities connected on both sides, which are required to compute the currents at each via segment.

$$\begin{bmatrix} \overline{\overline{h}}^{qq} & \overline{\overline{h}}^{qp} \\ \overline{\overline{h}}^{pq} & \overline{\overline{h}}^{pp} \end{bmatrix} = \begin{bmatrix} \overline{\overline{Z}}^{qq} - \overline{\overline{Z}}^{qp}\overline{\overline{Z}}^{pp^{-1}}\overline{\overline{Z}}^{pq} & \overline{\overline{Z}}^{qp}\overline{\overline{Z}}^{pp^{-1}} \\ \overline{\overline{Z}}^{pp^{-1}}\overline{\overline{Z}}^{pq} & \overline{\overline{Z}}^{pp^{-1}} \end{bmatrix}. \tag{6.9}$$

Note that $\overline{\overline{h}}^{pp}$ is the same as the parallel-plate admittance matrix $\overline{\overline{Y}}^{pp}$ with the boundary ports open, i.e., PMC boundary. As in Eq. (4.3), for the voltages and currents on the upper and lower via ports, the following relation holds: $\overline{V}^{u} - \overline{V}^{l} = \overline{V}^{p}$ and $\overline{I}^{u} = -\overline{I}^{l} = \overline{I}^{p}$. Therefore, the h-matrix can be expanded as:

$$\underbrace{\begin{bmatrix} \overline{\overline{h}}^{qq} & \overline{\overline{h}}^{qp} & -\overline{\overline{h}}^{qp} \\ \overline{\overline{h}}^{pq} & \overline{\overline{h}}^{pp} & -\overline{\overline{h}}^{pp} \\ -\overline{\overline{h}}^{pq} & -\overline{\overline{h}}^{pp} & \overline{\overline{h}}^{pp} \end{bmatrix}}_{\overline{\overline{h}}^{c}} \cdot \begin{bmatrix} \overline{I}^{q} \\ \overline{V}^{u} \\ \overline{V}^{l} \end{bmatrix} = \begin{bmatrix} \overline{V}^{q} \\ \overline{I}^{u} \\ \overline{I}^{l} \end{bmatrix}. \tag{6.10}$$

In $\overline{\overline{h}}^{c}$, the sub-matrix $\begin{bmatrix} \overline{\overline{h}}^{pp} & -\overline{\overline{h}}^{pp} \\ -\overline{\overline{h}}^{pp} & \overline{\overline{h}}^{pp} \end{bmatrix}$ is identical to the expanded parallel-plate admittance matrix in Eq. (4.3). Therefore, the via capacitance and stripline models can be incorporated according to Eq. (4.38). Consequently, the following h-parameter expression can be found to describe the second cavity including all the elements in Figure 6.30

$$\overline{\overline{h}}^{c} = \begin{bmatrix} \overline{\overline{h}}^{qq} & \overline{\overline{h}}^{qp} & -\overline{\overline{h}}^{qp} \\ \overline{\overline{h}}^{pq} & \overline{\overline{Y}}^{pp} + \overline{\overline{Y}}^{cu} + k^2 \overline{\overline{Y}}^{tl} & -\overline{\overline{Y}}^{pp} + (-k^2 - k)\overline{\overline{Y}}^{tl} \\ -\overline{\overline{h}}^{pq} & -\overline{\overline{Y}}^{pp} + (-k^2 - k)\overline{\overline{Y}}^{tl} & \overline{\overline{Y}}^{pp} + \overline{\overline{Y}}^{cl} + (k^2 + 2k + 1)\overline{\overline{Y}}^{tl} \end{bmatrix} . \tag{6.11}$$

Equation (6.11) describes the general network representation for an individual cavity. The coupling between the stripline and the boundary has been neglected.

For a single-cavity system, Eq. (6.11) is ready to be connected to source ports and terminations on the top and bottom sides of the cavity. The terminations, such as decoupling capacitors, are reduced using the segmentation technique. For that purpose, the h-parameters are transformed to Z-parameters, for which formulae to conduct segmentation are available. As a result, only the source ports, denoted as s, and the boundary ports q, as shown in Figure 6.30, remain in the final system matrix

$$\begin{bmatrix} \overline{\overline{Z}}^{qq} & \overline{\overline{Z}}^{qs} \\ \overline{\overline{Z}}^{sq} & \overline{\overline{Z}}^{ss} \end{bmatrix} \cdot \begin{bmatrix} \overline{I}^{q} \\ \overline{I}^{s} \end{bmatrix} = \begin{bmatrix} \overline{V}^{q} \\ \overline{V}^{s} \end{bmatrix} . \tag{6.12}$$

Since $\overline{I}^{q} = 0$ owing to the PMC boundary condition, the voltage distribution on the boundary ports $\overline{V}^{q}$ can be easily computed as

$$\left. \overline{V}^{q} \right|_{\overline{I}^{q}=0} = \overline{\overline{Z}}^{qs} \cdot \overline{I}^{s} . \tag{6.13}$$

The radiated far field is then calculated using the equivalence principle (Appendix A.10). In order to obtain the field distribution inside the cavity, the currents flowing on each via, including that connecting the terminations, must be known. The voltages on the upper and lower via ports $\overline{V}^{u}$ and $\overline{V}^{l}$ can be retrieved by storing an auxiliary matrix when applying segmentation [78]. The via currents are then obtained by using $\overline{I}^{p} = \overline{\overline{Y}}^{pp} \cdot \overline{V}^{p} = \overline{\overline{Y}}^{pp} \cdot \left(\overline{V}^{u} - \overline{V}^{l} \right)$, and the voltage distribution inside the cavity is obtained finally by CIM (Eq. (A9.5), Appendix A.9).

For multilayer systems, each individual cavity is described using Eq. (6.11). In principle, the cavities, including all the boundary ports, can be concatenated sequentially from top to bottom in one step to generate a transfer function between top-/bottom-layer source ports and the boundary ports of all the cavities. However, for the retrieval of voltages and currents on the intermediate via ports, as shown in Figure 6.29, it is required that auxiliary matrices be stored each time a cavity is appended. Hence, the amount of storage can be very large, especially for many layer systems. Alternatively, the current implementation follows a cavity-by-cavity approach, and, as depicted in Figure 6.30, each cavity is solved independently. The layers on its top and bottom sides are accounted for by two equivalent blocks, $\overline{\overline{Z}}^{u}_{eq}$ and $\overline{\overline{Z}}^{l}_{eq}$, which are obtained using the methods described in Chapter 4. The cavities within the equivalent blocks are represented using Eq. (4.38), without boundary ports. Consequently, the transfer function between the source ports and the boundary ports of only one intermediate cavity is obtained, and thus the voltage distribution on that cavity boundary. This process is repeated for every cavity to obtain the complete voltage distribution around the PCB sidewalls in order to calculate the far-field radiation. Here, since the boundary ports are only needed for the calculated cavity, the amount of storage is reduced in exchange for a slightly longer calculation time.

6.3.2. Application Example

The combined method has been applied to simulate several configurations. The goal of these analyses was to simultaneously obtain different metrics frequently used in SI, PI, and EMI applications such as S-parameters, Z-parameters, field distributions, radiation diagrams, and radiated power. The results have been compared against full-wave simulations. This section will explore one multilayer case including traces and ground vias. Additional examples can be found in [9],[17].

The structure considers a multilayer six-cavity board, with two single-ended links connected by four through-hole vias and eight ground vias shorting all the ground planes, as depicted in Figure 6.31. The traces are located in the third cavity and vertically centered (k = -0.5). Their width is 4 mil. The ground-signal via pitch is 40 mil. The lower via ends are assumed to be open and four ports are defined on top of the signal vias. It was assumed that all the reference planes are ground planes.

The transmission line model in the proposed method was obtained analytically, assuming a characteristic impedance of 49 Ω. The via capacitance values per cavity side were 26 fF for the inner layers and 29 fF for the most top and bottom layers [17]. No fringing fields from via ends were modeled in this example.

The S-parameters are plotted in Figure 6.32, which show good correlation between the proposed method and the full-wave analysis. Figure 6.33 shows the surface map of the electric field distribution inside the first cavity at 2.4 GHz, obtained by both the proposed method and the full-wave simulation. An incident power of 1 W applied at port 1 was used as excitation for both cases. The electric field amplitudes along the observation path for different layers are shown in Figure 6.34. A similar behavior is

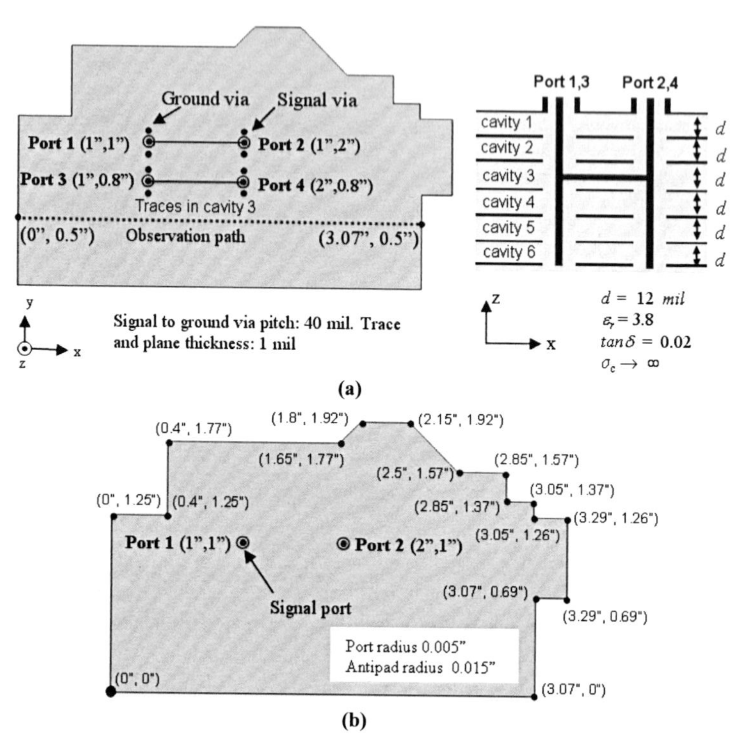

Figure 6.31 Description of the case study. (a). Irregular-shaped multilayer board with two links and eight ground vias. The ground vias are shorted to all seven ground planes. Dimensions are given in inches (1 mil = 0.001 inch $\approx 25.4 \cdot 10^{-6}$ m). (b) Plane and port position and dimensions.

predicted by the two techniques. The noise voltage is relatively large for the first two cavities due to the signal currents flowing on the via segments and through the trace, decreasing for the lower cavities at this frequency. Figure 6.35 provides the radiation diagram of the electric far-field at 2.4 GHz and a distance of 10 meters. The radiated power is shown in Figure 6.36; good agreement is observed up to 10 GHz. Beyond that, the resonances predicted by the proposed method are more noticeable than by the FEM simulation. This could be due to the fact that the proposed method assumes PMC boundary, and hence, the power leakage through the cavity edges is neglected, resulting in a higher Q factor of the resonances. Further investigations are necessary to address this issue.

For comparison, the computation time for the multilayer case was about 98 seconds for 200 frequencies, on a 3.0-GHz CPU, 4-GB RAM PC. The full-wave analysis needed a CPU time of 22 hours 53 minutes to solve the same problem on the same computer.

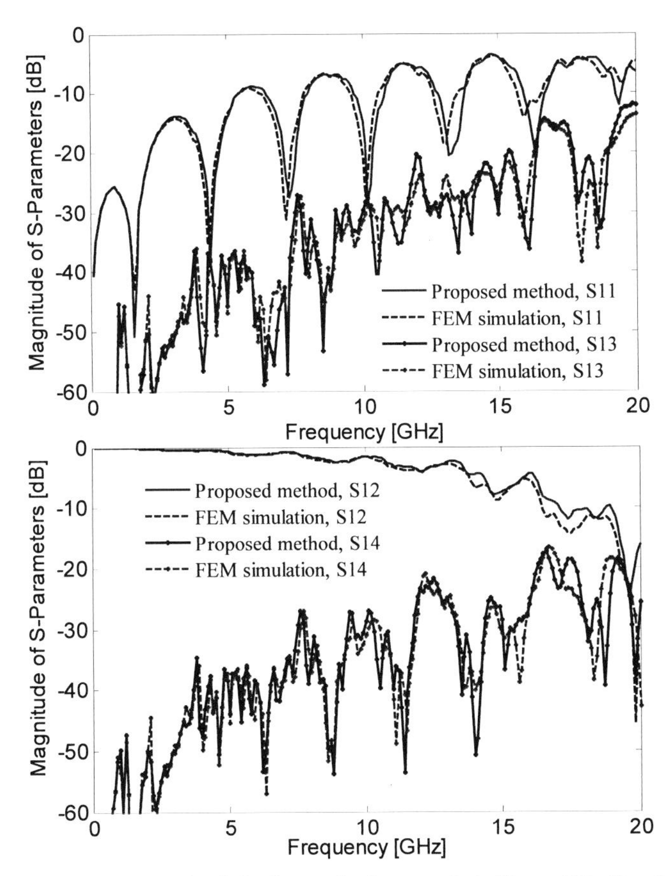

Figure 6.32 *S*-parameters at the defined ports for the example in Figure 6.31. The plots show the results obtained with the proposed method and a full-wave analysis.

The proposed method is however 2D-based and some of the full-wave effects cannot be captured in the current implementation. Also, the method becomes less accurate when the radiation increases and may not be suitable for efficient radiating structures. The power leakage through cavity boundaries and cross coupling between cavities have been neglected. This might cause the deviation with respect to the full-wave analysis, especially for the multilayer case at higher frequencies. Also, elements located in close proximity to the board edge, as well as slot and split planes cannot be easily included.

Figure 6.32 to Figure 6.36 illustrate a diverse set of outputs that allow analyses of multilayer PCB from different perspectives. The method can provide results as S-parameters, Z-parameters, field distributions between power planes, radiation diagrams,

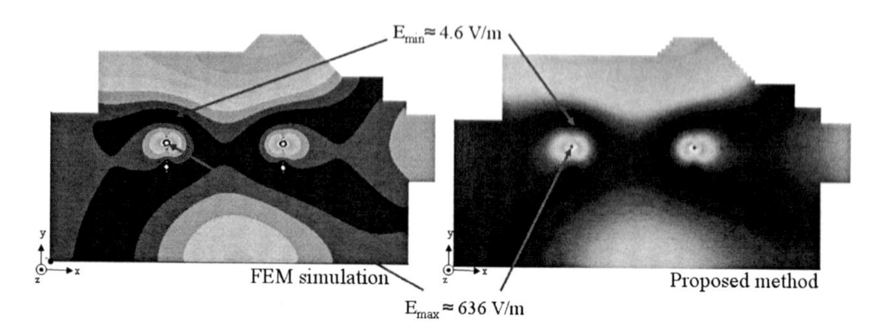

Figure 6.33 Electric field distribution for the first cavity at 2.4 GHz, obtained with the proposed method and a full-wave simulation.

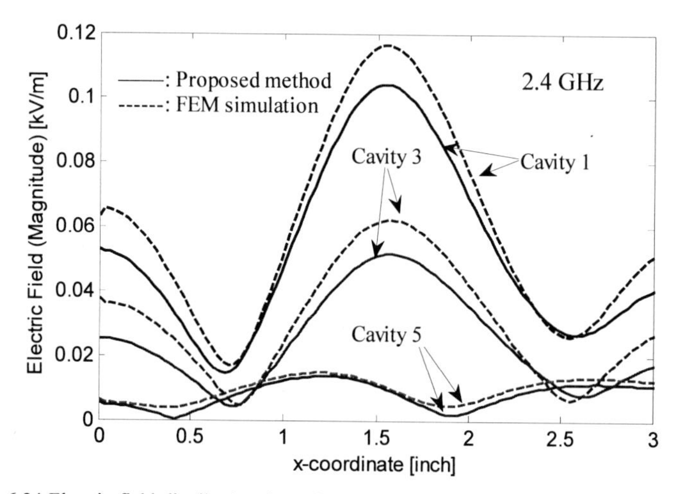

Figure 6.34 Electric field distribution (complex field amplitude) along the observation path (Figure 6.31) inside cavity 1, 3, and 5 at 2.4 GHz, obtained by both the proposed method and a full-wave simulation.

and radiated power that are useful to SI, PI, and EMC applications. In principle, the proposed method can be applied up to the frequency where the behavior of the via structure and the parallel planes show the effect of higher order propagating modes. Based on previous results, it is projected that the proposed method can be extended to approximately 40 GHz.

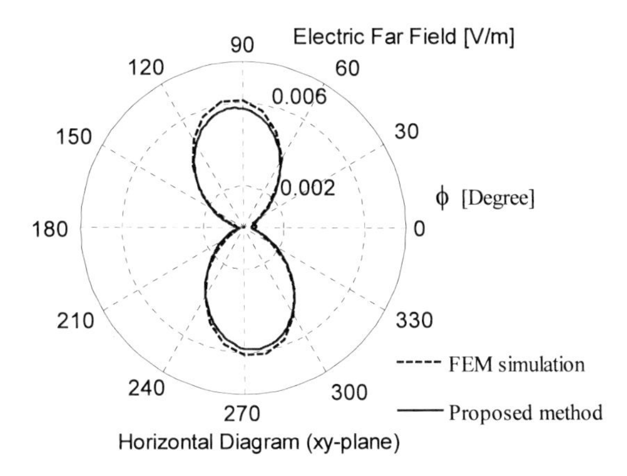

Figure 6.35 Electric far-field radiation diagrams at 2.4 GHz and a distance of 10 meter, simulated with the proposed method and a full-wave analysis.

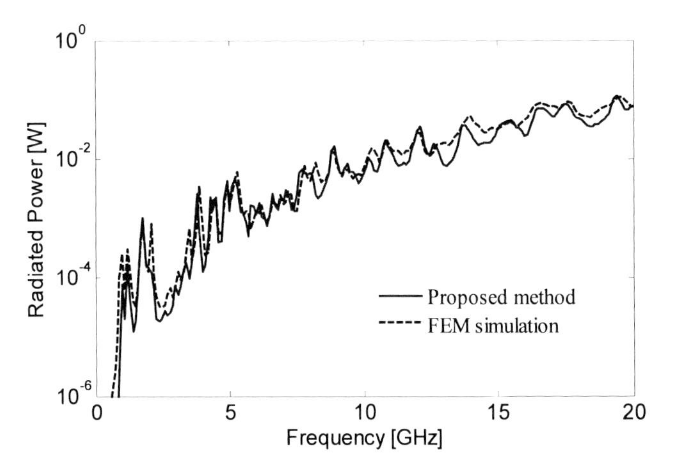

Figure 6.36 Radiated power simulated with the proposed method and a full-wave analysis.

7. Conclusions and Outlook

This thesis has presented efficient semi-analytical models for the electrical behavior of vias and traces found in multilayer substrates, and proposed a simulation framework for their utilization, based on microwave network parameters. Results comparable to those from general-purpose numerical techniques and measurements have been obtained in the GHz range. Several application scenarios have been evaluated from signal integrity, power integrity, and electromagnetic compatibility perspectives. These models have proven to be useful in developing an understanding of the physics of the problem, and for rapid design studies and prototyping. Their high numerical efficiency, from two to over three orders of magnitude higher than full-wave methods, makes the models suitable for simulation of complex cases. Structures with hundreds of power, ground and signal vias, over ten cavities, several coupled and uncoupled traces at different levels, and lumped elements such as decoupling capacitors have been simulated and analyzed. In summary, the main scientific contributions presented in this thesis are:

- Semi-analytical models for vias and traces in multilayer substrates have been proposed. The existent "physics-based" via model has been extended to consider traces by applying a modal decomposition technique (Section 4.6).
- These models have been formulated in terms of microwave network parameters and incorporated into a generalized and semi-automated simulation method that can also consider power vias, ground vias, and arbitrary reference plane assignments (Sections 4.2, 4.4-4.7).
- Methods to analytically compute the parallel-plate impedance have been explored and a hybrid method combining the cavity resonator model and the radial waveguide approach has been proposed (Section 4.3).
- An extensive and thorough validation of the models has been carried out considering several different structures of practical interest: multilayer configurations with power, ground and signal vias (Sections 5.1-5.4), single-ended links (Section 5.2), mixed reference stackups (Section 5.2.4), structures with buried and blind vias (Section 5.2.5), differential links (Section 5.3), and via arrays (Section 5.4). The models have been extensively validated with other numerical techniques (FEM and FIT) and measurements. Model advantages,

limitations, and efficiency aspects have been addressed in the analysis (Sections 5.5-5.6).

- It has been shown that these models are suitable for analysis of structures with realistic complexity and for investigation of diverse aspects such as the performance of differential links (Section 6.1.1), stub resonances (Section 6.1.2), and mode conversion mechanisms (Section 6.1.3).
- The applicability of the method for co-simulation and co-analysis of signal and power integrity has been demonstrated (Section 6.2).
- A combined method for efficient simulation of arbitrary shaped reference planes considering signal integrity, power integrity, and radiated emissions has been proposed and validated (Section 6.3).

The work performed indicates that the proposed models and the simulation framework are versatile and flexible enough to allow the pre-layout simulation and optimization of complex interconnect systems in multilayer substrates. The discussed approach constitutes a promising alternative to assist real system-level analyses of interconnects in high-speed electronic systems. The studies show that the models can provide good accuracy up to 20 GHz and fair up to 40 GHz.

At the present time, further improvement of the models and investigation of their applicability to more dense and complex systems is necessary. Recommendations for future work comprise the extension of the models to include the effects of higher order propagating and evanescent modes in the cavities. The coupling through near fields of vias is possible for very dense arrays, and the scattering among via barrels has not yet been considered in these models. Moreover, the excitation of non-uniform cylindrical modes and the modeling of electrically large ports need to be further investigated. It is important to determine the geometries and frequency range for which these effects begin to significantly impact the results.

Refinements of the simulation method are also recommended. Fast methods to calculate the via-to-plane capacitances for arbitrary pad stacks are still required. The effect of arbitrary pads and trace connections on the field near to vias also needs more investigation. In addition, the impact of diverse sources of mode conversion between the parallel-plate modes and the transmission lines, and other coupling mechanisms, such as inter-cavity trace coupling, will require further analysis. The modeling of irregular planes, slot lines and perforations has shown to be possible with the utilization of 2D customized methods such as the CIM. However, efficient techniques for modeling of the inter-cavity coupling in multilayer structures through plate discontinuities and board edges have not been thoroughly addressed. Broadband models for the portion of the substrate not enclosed by reference planes and additional interconnects such as solder balls or connectors are still necessary to extend the applicability of the method.

Predictions made with the radial waveguide approach can suffer from passivity violations. Therefore, alternatives to ensure causality and passivity of the results are expected to be required for application of the models to more complex configurations and at higher frequencies.

The work on model-to-hardware correlation has revealed other important challenges. Process variations and tolerances can play an important role. The impact of the calibration and signal launches are topics where more research should be conducted. More advanced loss models can improve the accuracy of the results over a broader bandwidth. The mapping of frequency dependencies of material and model parameters is another issue for further research.

Finally, the available code can be optimized for speed. Different alternatives to compute and combine the results should be explored in order to improve the overall efficiency. Lower level code implementation, parallelization, and more efficient memory management are tasks that can be addressed in the future. The evaluation of the proposed approach with respect to other hybrid approaches (e.g. [65],[77]) is also suggested as future work.

A. Mathematical Appendix

A.1. Bandwidth of Digital Signals

The bandwidth of a digital signal for terminated circuits can be approximated from a periodic and even trapezoidal pulse train [39],[45],[50] (Figure A.1).

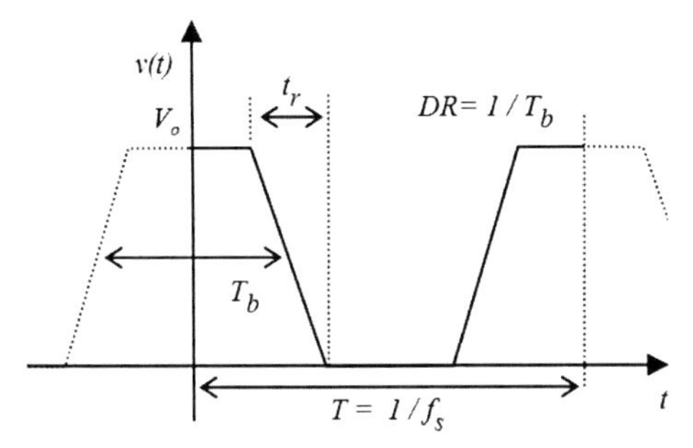

Figure A.1 Idealized digital signal represented by an even trapezoidal pulse train.

The Fourier series to represent the signal is [39]

$$v(t) = a_0 + \sum_{n=1}^{N} a_n \cos\left(\frac{2\pi n}{T} t\right), \tag{A1.1}$$

with the DC value

$$a_0 = \frac{2}{T}\int_0^{T/2} v(t)\,dt = V_0 \frac{T_b}{T} \ , \tag{A1.1a}$$

and the coefficients

$$a_n = \frac{4}{T}\int_0^{T/2} v(t)\cos\left(\frac{2\pi n}{T}\right) dt = 2 \cdot a_0 \operatorname{sin c}\left(\frac{n\pi T_b}{T}\right) \operatorname{sin c}\left(\frac{n\pi t_r}{T}\right) \quad \text{for} \quad n = 1,2,3\ldots, \tag{A1.1b}$$

where $\operatorname{sin}c(x)=\sin(x)/x$. A large enough number of harmonics N is required in order to reconstruct the signal with good accuracy (Figure A.2).

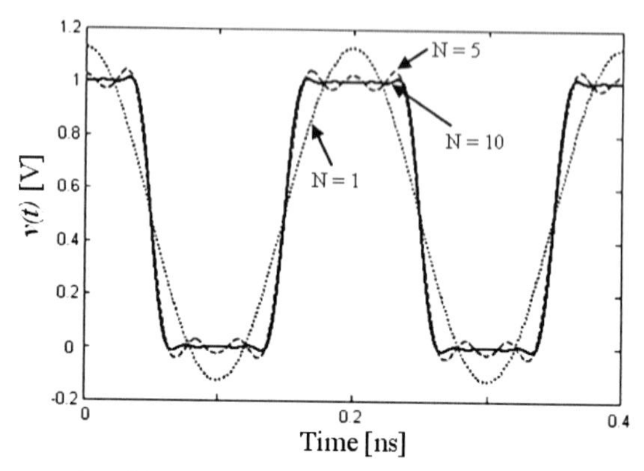

Figure A.2 Representation of the trapezoidal signal with a different number of harmonics, for a data rate of 10 Gb/s ($T = 20$ ps) a rise time of $t_r = 0.1T$.

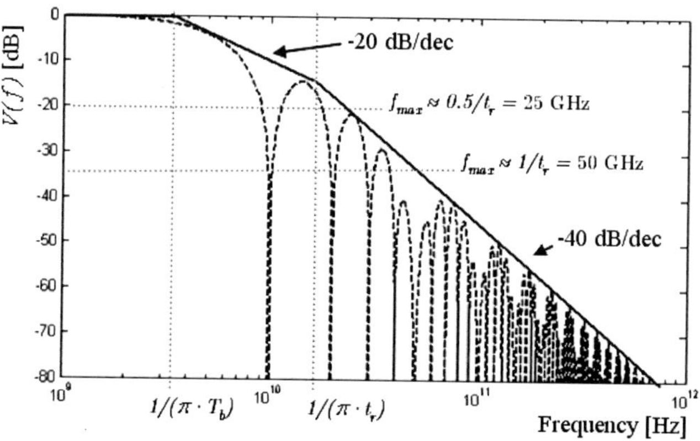

Figure A.3 Envelope of the spectrum of a trapezoidal signal for a data rate of 10 Gb/s ($T = 20$ ps) and a rise time of $t_r = 0.1T$.

The envelope of the spectrum in dB can be obtained from the Fourier series by substituting $f = n/T$ in Eq. (A1.1) [45]

$$20 \cdot \log\left|V\left(f\right)\right| = 20 \cdot \log\left|2 \cdot V_0 \frac{T_b}{T}\right| + 20 \cdot \log\left|\sin c\left(\pi T_b f\right)\right| + 20 \cdot \log\left|\sin c\left(\pi t_r f\right)\right|. \qquad \text{(A1.2)}$$

The spectrum is plotted in Figure A.3 for a signal with a data rate of 10 Gb/s. The rules-of-thumb for the maximum frequency content of the signal truncate the spectrum at approximately -20 dB and -35 dB for $0.5/t_r$ and $1/t_r$, respectively.

A.2. Microwave Network Parameters

The microwave network theory describes systems in terms of equivalent voltages and currents defined between terminal pairs, called ports. For an N-port system, the total voltages (V) and currents (I) —at the reference planes where the ports are defined— are given by [124],[155] (Figure A.4)

$$V_n = V_n^+ + V_n^- \,,\; I_n = I_n^+ - I_n^- \,. \tag{A2.1}$$

The superscripts + and – denote incident and reflected quantities, respectively, which can be expressed into different forms. Impedance parameters (Z-parameters) are defined as

$$\overline{V} = \overline{\overline{Z}} \cdot \overline{I} \,, \tag{A2.2}$$

with

$$\overline{V} = [V_1, V_2, ... V_N]^T \,,\; \overline{I} = [I_1, I_2, ... I_N]^T \,, \tag{A2.2a}$$

and

$$Z_{ij} = \left. \frac{V_i}{I_j} \right|_{I_k = 0, k \neq j} \,. \tag{A2.2b}$$

The admittance (Y) matrix, which is the inverse of the Z matrix, is defined as

$$\overline{I} = \overline{\overline{Y}} \cdot \overline{V} \,, \tag{A2.3}$$

with

$$Y_{ij} = \left. \frac{I_i}{V_j} \right|_{V_k = 0, k \neq j} \,. \tag{A2.3a}$$

If the N-ports are separated as input and output ports (or upper an lower ports for the case of multilayer structures), the chain matrix ($ABCD$-parameters) is given by

$$\begin{bmatrix} \overline{V}_{in} \\ \overline{I}_{in} \end{bmatrix} = \begin{bmatrix} \overline{\overline{A}} & \overline{\overline{B}} \\ \overline{\overline{C}} & \overline{\overline{D}} \end{bmatrix} \cdot \begin{bmatrix} \overline{V}_{out} \\ \overline{I}_{out} \end{bmatrix} . \tag{A2.4}$$

where

$$\overline{V}_{in} = [V_1, V_2, ... V_n]^T, \qquad \overline{V}_{out} = [V_{n+1}, V_{n+2}, ... V_N]^T \,, \tag{A2.4a}$$

$$\overline{I}_{in} = [I_1, I_2, ... I_n]^T, \qquad \overline{I}_{out} = [I_{n+1}, I_{n+2}, ... I_N]^T \,, \tag{A2.4b}$$

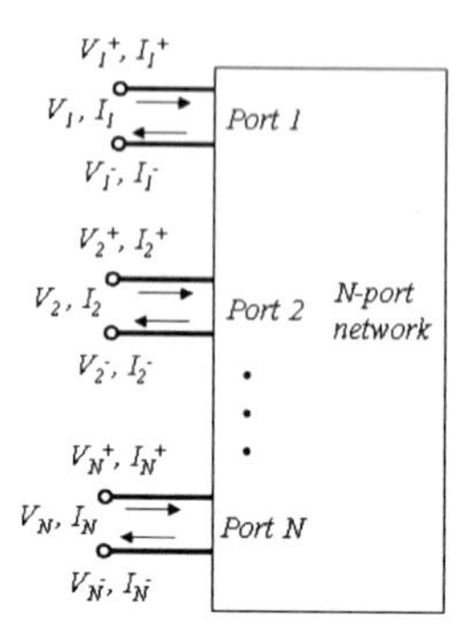

Figure A.4 Illustration of an N-port network.

The *ABCD* matrix allows the cascade connection of multiple networks by simple matrix multiplication.

Scattering parameters (*S*) are widely used for signal integrity applications since they are often more intuitive and are directly related to important performance metrics such as reflection, transmission, and crosstalk. The scattering matrix is defined in terms of incident and reflected waves [39]

$$\overline{V}^- = \overline{\overline{S}} \cdot \overline{V}^+ . \tag{A2.5}$$

The incident and reflected wave vectors are

$$\overline{V}^+ = \left[V_1^+, V_2^+, \ldots V_N^+\right]^T , \tag{A2.5a}$$

$$\overline{V}^- = \left[V_1^-, V_2^-, \ldots V_N^-\right]^T , \tag{A2.5b}$$

and the matrices entries are given by

$$S_{ij} = \left.\frac{V_i^-}{V_j^+}\right|_{V_k^+ = 0, k \neq j} . \tag{A2.5c}$$

The generalized *S*-parameters consider the case that different ports may have different reference impedances Z_0, and the *S*-parameters are redefined as [155]

$$\bar{b} = \overline{\overline{S}} \cdot \bar{a} , \tag{A2.6}$$

$$\bar{a} = \left[V_1^+ / \sqrt{Z_{01}}, V_2^+ / \sqrt{Z_{02}}, \ldots V_N^+ / \sqrt{Z_{0N}}\right]^T , \tag{A2.6a}$$

$$\bar{b} = \left[V_1^- / \sqrt{Z_{01}}, V_2^- / \sqrt{Z_{02}}, \ldots V_N^- / \sqrt{Z_{0N}} \right]^T . \quad \text{(A2.6b)}$$

The port voltages and currents for the *n-th* port are given by

$$V_n = V_n^+ + V_n^- = \sqrt{Z_{0n}} (a_n + b_n) , \quad \text{(A2.6c)}$$

$$I_n = I_n^+ - I_n^- = \frac{1}{\sqrt{Z_{0n}}} (a_n - b_n) . \quad \text{(A2.6d)}$$

With the S-matrix entries

$$S_{ij} = \left. \frac{b_i}{a_j} \right|_{a_k = 0, k \neq j} . \quad \text{(A2.6e)}$$

The more convenient parameter definition depends on the topology of the network and the metrics to be evaluated. Other forms are also commonly found including h-parameters (see section 6.3), and transmission scattering matrices [124],[151].

Since all parameters are derived from the same definition, a microwave network can be transformed to other parameter forms, with exception of the cases where singularities arise. The transformations used in this work, generalized for N ports, are detailed as follows.

Between Z and Y- parameters, it holds that

$$\bar{\bar{Y}} \Leftrightarrow \bar{\bar{Z}}^{-1} . \quad \text{(A2.7)}$$

For Z- and $ABCD$-parameters, the following identities can be used [151],[155]

$$\begin{bmatrix} \bar{\bar{Z}}_{11} & \bar{\bar{Z}}_{12} \\ \bar{\bar{Z}}_{21} & \bar{\bar{Z}}_{22} \end{bmatrix} = \begin{bmatrix} \bar{\bar{A}}\bar{\bar{C}}^{-1} & \bar{\bar{A}}\bar{\bar{C}}^{-1}\bar{\bar{D}} - \bar{\bar{B}} \\ \bar{\bar{C}}^{-1} & \bar{\bar{C}}^{-1}\bar{\bar{D}} \end{bmatrix} , \quad \text{(A2.8a)}$$

$$\begin{bmatrix} \bar{\bar{A}} & \bar{\bar{B}} \\ \bar{\bar{C}} & \bar{\bar{D}} \end{bmatrix} = \begin{bmatrix} \bar{\bar{Z}}_{11}\bar{\bar{Z}}_{21}^{-1} & \bar{\bar{Z}}_{11}\bar{\bar{Z}}_{21}^{-1}\bar{\bar{Z}}_{22} - \bar{\bar{Z}}_{12} \\ \bar{\bar{Z}}_{21}^{-1} & \bar{\bar{Z}}_{21}^{-1}\bar{\bar{Z}}_{22} \end{bmatrix} . \quad \text{(A2.8b)}$$

In this work, it is assumed that the matrices are reciprocal, and the Z-matrix corresponds to the definition

$$\begin{bmatrix} \bar{V}_{in} \\ \bar{V}_{out} \end{bmatrix} = \begin{bmatrix} \bar{\bar{Z}}_{11} & \bar{\bar{Z}}_{12} \\ \bar{\bar{Z}}_{21} & \bar{\bar{Z}}_{22} \end{bmatrix} \cdot \begin{bmatrix} \bar{I}_{in} \\ \bar{I}_{out} \end{bmatrix} . \quad \text{(A2.9)}$$

Similarly, between Y- and $ABCD$ parameters it can be written [151],[155]

$$\begin{bmatrix} \overline{\overline{Y}}_{11} & \overline{\overline{Y}}_{12} \\ \overline{\overline{Y}}_{21} & \overline{\overline{Y}}_{22} \end{bmatrix} = \begin{bmatrix} \overline{\overline{D}}\,\overline{\overline{B}}^{-1} & -\overline{\overline{A}}\,\overline{\overline{B}}^{-1}\overline{\overline{D}} + \overline{\overline{C}} \\ -\overline{\overline{B}}^{-1} & \overline{\overline{B}}^{-1}\overline{\overline{A}} \end{bmatrix}, \tag{A2.10a}$$

$$\begin{bmatrix} \overline{\overline{A}} & \overline{\overline{B}} \\ \overline{\overline{C}} & \overline{\overline{D}} \end{bmatrix} = \begin{bmatrix} -\overline{\overline{Y}}_{21}^{\,-1}\overline{\overline{Y}}_{22} & -\overline{\overline{Y}}_{21}^{\,-1} \\ -\overline{\overline{Y}}_{11}\overline{\overline{Y}}_{21}^{\,-1}\overline{\overline{Y}}_{22} + \overline{\overline{Y}}_{12} & -\overline{\overline{Y}}_{11}\overline{\overline{Y}}_{21}^{\,-1} \end{bmatrix}. \tag{A2.10b}$$

Assuming that all the N-ports have the same reference impedance Z_0, the transformations related to S-parameters are given by [39],[151],[155]

$$\overline{\overline{S}} = \left[\overline{\overline{Z}} - Z_0\overline{\overline{E}}\right] \cdot \left[\overline{\overline{Z}} + Z_0\overline{\overline{E}}\right]^{-1}, \tag{A2.11a}$$

$$\overline{\overline{Z}} = Z_0 \cdot \left[\overline{\overline{E}} - \overline{\overline{S}}\right]^{-1} \cdot \left[\overline{\overline{E}} + \overline{\overline{S}}\right], \tag{A2.11b}$$

and

$$\overline{\overline{S}} = \left[\frac{1}{Z_0}\cdot\overline{\overline{E}} + \overline{\overline{Y}}\right]^{-1} \cdot \left[\frac{1}{Z_0}\cdot\overline{\overline{E}} - \overline{\overline{Y}}\right], \tag{A2.12a}$$

$$\overline{\overline{Y}} = \frac{1}{Z_0} \cdot \left[\overline{\overline{E}} + \overline{\overline{S}}\right]^{-1} \cdot \left[\overline{\overline{E}} - \overline{\overline{S}}\right]. \tag{A2.12b}$$

$\overline{\overline{E}}$ denotes the *N-times-N* identity matrix.

A.3. Via Model Matrix Expansion

Recalling Eqs. (4.1) and (4.2)

$$\overline{V} = \overline{\overline{Z}}^{pp} \cdot \overline{I}, \tag{4.1}$$

$$V_i = V_i^u - V_i^l,\ I_i = I_i^u = -I_i^l, \tag{4.2}$$

the parallel-plate impedance matrix is expanded to upper and lower ports by using the relations in Eq. (4.2), as Y-parameters

$$\overline{I} = \overline{\overline{Y}}^{pp} \cdot \overline{V}, \tag{A3.1}$$

$$\overline{I}^u = \left(\overline{V}^u - \overline{V}^l\right) \cdot \overline{\overline{Y}}^{pp} = \overline{\overline{Y}}^{pp} \cdot \overline{V}^u - \overline{\overline{Y}}^{pp} \cdot \overline{V}^l, \tag{A3.1a}$$

$$\overline{I}^l = -\left(\overline{V}^u - \overline{V}^l\right) \cdot \overline{\overline{Y}}^{pp} = -\overline{\overline{Y}}^{pp} \cdot \overline{V}^u + \overline{\overline{Y}}^{pp} \cdot \overline{V}^l. \tag{A3.1b}$$

Therefore, the expanded admittance matrix can be written as in Section 4.2

$$\begin{bmatrix} \overline{I}^u \\ \overline{I}^l \end{bmatrix} = \begin{bmatrix} \overline{\overline{Y}}^{pp} & -\overline{\overline{Y}}^{pp} \\ -\overline{\overline{Y}}^{pp} & \overline{\overline{Y}}^{pp} \end{bmatrix} \cdot \begin{bmatrix} \overline{V}^u \\ \overline{V}^l \end{bmatrix}. \tag{4.3}$$

The expression in Eq. (4.3) is singular for Z-parameters [124]. The formulation as Y-parameters allows the inclusion of the via capacitances easily, since the topology corresponds to a π-network, which requires just the addition of the self-admittances of the parallel-plate model with the capacitance matrices, as in Eq. (4.4).

The matrix in Eq. (4.3) can be however transformed to other parameter forms such as $ABCD$ [16]

$$\begin{bmatrix} \overline{V}^u \\ \overline{I}^u \end{bmatrix} = \begin{bmatrix} \overline{\overline{E}} & \overline{\overline{Z}}^{pp} \\ \overline{\overline{0}} & \overline{\overline{E}} \end{bmatrix} \cdot \begin{bmatrix} \overline{V}^l \\ \overline{I}^l \end{bmatrix}, \tag{A3.2}$$

with $\overline{\overline{E}}$ the identity matrix. The via-to-plane capacitances of each cavity side can be written as chain matrices as

$$\begin{bmatrix} \overline{V}^u \\ \overline{I}^u \end{bmatrix} = \begin{bmatrix} \overline{\overline{E}} & 0 \\ \overline{\overline{Y}}^{cu/l} & \overline{\overline{E}} \end{bmatrix} \cdot \begin{bmatrix} \overline{V}^l \\ \overline{I}^l \end{bmatrix}. \tag{A3.3}$$

$\overline{\overline{Y}}^{cu/l}$ is a diagonal matrix with entries $Y_i^c = 1/Z_i^c = j\omega C_i^v$, corresponding to the capacitance values. The $ABCD$ formulation is advantageous to simulate regular configurations since the concatenation procedure can be done by matrix multiplication [16].

A.4. Derivation of the Radial Waveguide Formula to Compute Z^{pp}

The formulation to compute Z^{pp} with the assumption of infinite planes is derived from the transmission line equations [156],[157]

$$\frac{\partial V}{\partial \rho} = -j\underline{k}Z \cdot I\,, \tag{A4.1a}$$

$$\frac{\partial I}{\partial \rho} = -j\underline{k}Y \cdot V\,, \tag{A4.1b}$$

whose solution for cylindrical waves has the general form (see geometry in Figure 4.5)

$$V(\omega) = A \cdot H_0^{(2)}(\underline{k}\rho) \cdot I(\omega)\,, \tag{A4.2}$$

where $H_0^{(2)}$ is the Hankel function of order 0 and second kind. The derivative of Eq. (A4.2) is given by

$$\frac{\partial V(\omega)}{\partial \rho} = A \cdot \underline{k} \cdot H_1^{(2)}(\underline{k}\rho) \cdot I(\omega)\,. \tag{A4.3}$$

To find the constant A, Eq. (A4.3) and Eq. (A4.1a) are evaluated at the perimeter of the port I, with $\rho = \rho_o$ [102]

$$\frac{\partial V_i(\omega)}{\partial \rho} = A \cdot \underline{k} \cdot H_1^{(2)}(\underline{k}\rho_0) \cdot I_i(\omega) = -j\underline{k}Z \cdot I_i(\omega)\,, \tag{A4.4}$$

$$A = \frac{-jZ}{H_1^{(2)}(\underline{k}\rho_0)}\,. \tag{A4.5}$$

The self-impedance is then given by

$$\frac{V_i(\omega)}{I_i(\omega)} = A \cdot H_0^{(2)}(\underline{k}\rho_{ij}) = -jZ\frac{H_0^{(2)}(\underline{k}\rho_o)}{H_1^{(2)}(\underline{k}\rho_o)}\,. \tag{A4.6}$$

Under the assumption that only the TEM mode is propagating, the cavity is treated as a radial transmission line [85],[156],[157], where

$$V = \mathcal{E} \cdot d\,, \tag{A4.7a}$$

$$I = \mathcal{H} \cdot 2\pi\rho_0\,, \tag{A4.7b}$$

and

$$Z = \frac{V}{I} = \frac{-\mathcal{E} \cdot d}{\mathcal{H} \cdot 2\pi\rho_0} = -\frac{\eta \cdot d}{2\pi\rho_0} = \frac{\omega \cdot \mu \cdot d}{\underline{k}2\pi\rho_0}\,. \tag{A4.8}$$

For n-ports with radius ρ_0 the self-impedances are given by

$$Z_{ii}^{rw}(\omega) = \frac{V_i(\omega)}{I_i(\omega)} = \frac{j\eta d}{2\pi\rho_0 H_1^{(2)}(\underline{k}\rho_0)} \cdot H_0^{(2)}(\underline{k}\rho_o)\,. \tag{A4.9}$$

The transfer impedances are calculated as the induced voltage at the port i due to the current at the port j, neglecting the finite size of the port j [102]

$$Z_{ij}^{rw}(\omega) = \frac{V_i(\omega)}{I_j(\omega)} = \frac{j\eta d}{2\pi\rho_0 H_1^{(2)}(\underline{k}\rho_0)} \cdot H_0^{(2)}(\underline{k}\rho_{ij}), \qquad \text{for } i \neq j, \tag{A4.10}$$

with ρ_{ij} the radial distance between the ports.

According to the formulation presented by Chada *et al.* in [110], Eq. (A4.10) can be extended to consider the finite size of the ports for transfer impedances. The expression can be written as

$$Z_{ij}^{rw}(\omega) = \frac{j\omega\mu d}{2\pi\rho_0 \underline{k} H_1^{(2)}(\underline{k}\rho_0)} \cdot H_0^{(2)}(\underline{k}\rho_{ij}) J_0(\underline{k}\rho_{0j}), \qquad \text{for } i \neq j, \tag{A4.11}$$

where ρ_{0j} is the size of the port j.

A.5. Derivation of the Modal Transformation Matrices

The modal decomposition method presented in [79] is a solution to diagonalize the multiconductor transmission line equations (Eqs. (4.23)-(4.24)) for the parallel-plate modes and the transmission line modes associated with traces. Assuming perfect conductors and a homogeneous dielectric, the following identities are valid [45],[121],[122]

$$\overline{\overline{C}} \cdot \overline{\overline{L}} = \mu\varepsilon \overline{\overline{E}}, \tag{A5.1a}$$

$$\overline{\overline{G}} \cdot \overline{\overline{L}} = \mu\sigma \overline{\overline{E}}. \tag{A5.1b}$$

with $\overline{\overline{E}}$ denoting the identity matrix, μ the permeability, ε the permittivity and σ the conductivity of the dielectric material. The matrices C, L, G, refer to the per unit length (p.u.l) capacitance, inductance, and conductance of the multiconductor transmission lines, respectively.

Transformation matrices $\overline{\overline{T_v}}$ and $\overline{\overline{T_i}}$ that diagonalize the per unit length (p.u.l.) inductance matrix must fulfill the following condition

$$\overline{\overline{T_v}}^{-1} \overline{\overline{L}}\,\overline{\overline{T_i}} = \overline{\overline{L_m}} = \begin{pmatrix} \overline{\overline{L_{pp}}} & 0 \\ 0 & \overline{\overline{L_{tl}}} \end{pmatrix}. \tag{A5.2}$$

Equation (A5.2) diagonalizes the MTL equations, since in combination with Eqs. (A5.1a-A5.1b) it holds that [45]

$$\overline{\overline{L}}_m^{-1} = \left(\overline{\overline{T}}_v^{-1}\overline{\overline{L}}\,\overline{\overline{T}}_i\right)^{-1} = \overline{\overline{T}}_i^{-1}\overline{\overline{L}}^{-1}\overline{\overline{T}}_v = \frac{1}{\mu\varepsilon}\overline{\overline{T}}_i^{-1}\overline{\overline{C}}\,\overline{\overline{T}}_v, \tag{A5.3a}$$

$$\overline{\overline{L}}_m^{-1} = \frac{1}{\mu\sigma}\overline{\overline{T}}_i^{-1}\overline{\overline{G}}\,\overline{\overline{T}}_v. \tag{A5.3b}$$

One via-trace transition is considered next, setting the lower plane of a cavity as the reference potential (see definitions in Figure 4.13). The currents associated with the parallel-plate modes flow through the upper plane and return through the lower. The return currents of traces are distributed in different proportions to the two planes, depending on the position of the trace. The transformation matrix has then the form

$$\begin{pmatrix} I_i^u \\ I_i^s \end{pmatrix} = \overline{\overline{T}}_i \cdot \overline{I}_m = \begin{pmatrix} 1 & k_a \\ 0 & k_b \end{pmatrix}\begin{pmatrix} I_i^{pp} \\ I_i^{tl} \end{pmatrix}. \tag{A5.4a}$$

Similarly, the voltages between the planes are distributed by the factors k_c and k_d [45]

$$\begin{pmatrix} \phi_i^u - \phi_i^l \\ \phi_i^s - \phi_i^l \end{pmatrix} = \overline{\overline{T}}_v \cdot \overline{V}_m = \begin{pmatrix} k_c & 0 \\ k_d & 1 \end{pmatrix}\begin{pmatrix} V_i^{pp} \\ V_i^{tl} \end{pmatrix}. \tag{A5.4b}$$

If Eqs. (A5.2) and (A5.4a-A5.4b) are used, it is found that

$$\overline{\overline{T}}_v^{-1}\overline{\overline{L}}_m\overline{\overline{T}}_i = \begin{pmatrix} 1/k_c & 0 \\ -k_d/k_c & 1 \end{pmatrix}\begin{pmatrix} L_{pp} & L_{sp} \\ L_{ps} & L_{ss} \end{pmatrix}\begin{pmatrix} 1 & k_a \\ 0 & k_b \end{pmatrix}$$

$$= \begin{pmatrix} \dfrac{L_{pp}}{k_c} & \dfrac{k_a}{k_c}L_{pp} + \dfrac{k_b}{k_c}L_{sp} \\ -\dfrac{k_d}{k_c}L_{pp} + L_{sp} & -\dfrac{k_a k_d}{k_c}L_{pp} + k_a L_{sp} - \dfrac{k_d k_b}{k_c}L_{sp} + k_b L_{ss} \end{pmatrix} = \begin{pmatrix} L_{pp} & 0 \\ 0 & L_{tl} \end{pmatrix}, \tag{A5.5}$$

where L_{pp} is the self-inductance of the top plane, L_{sp}/L_{ps} the mutual inductances between the trace and the top plane, and L_{ss} the self-inductance of the trace. L_{pp} stands for the total inductance of the parallel plates and L_{tl} for the inductance of the trace. The matrix in Eq. (A5.5) is diagonalized if

$$k_a = -\frac{L_{sp}}{L_{pp}}k_b, \tag{A5.6a}$$

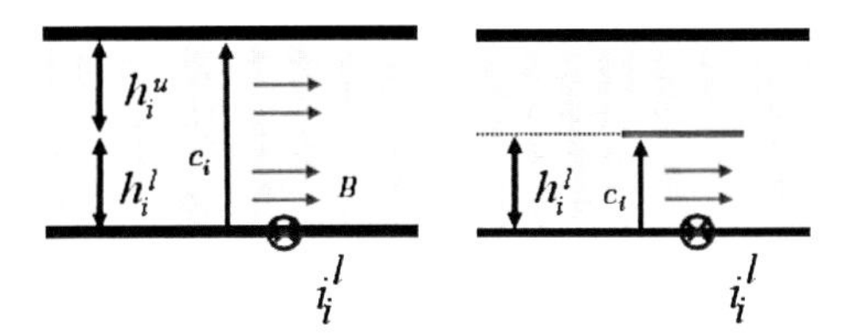

Figure A.5 Sketch for computation of the per-unit-length inductance.

$$k_d = \frac{L_{sp}}{L_{pp}} k_c . \tag{A5.6b}$$

The per-unit-length inductances Lsp and Lpp can be analytically estimated by neglecting the fringing fields (i.e. neglecting the trace thickness) as [122]

$$L_{ij} = \left. \frac{\psi_i}{I_j} \right|_{I_k=0, k\neq j} = \frac{\int_{c_i} B \cdot dh}{I_j} . \tag{A5.7}$$

with B the magnetic flux density and I_j the current at the port j. Therefore, it holds

$$L_{sp} = \frac{B(h_i^{\,l})}{i_i^l} , \tag{A5.7a}$$

$$L_{pp} = \frac{B(h_i^{\,l} + h_i^{\,u})}{i_i^l} . \tag{A5.7b}$$

Figure A.5 illustrates the field picture used to calculate the inductances. Since the total current flowing through the plane assigned as reference is the same in both expressions of Eq (A5.7), the factor k can be introduced

$$k_i = -\frac{L_{sp}}{L_{pp}} = -\frac{h_i^{\,l}}{h_i^{\,l} + h_i^{\,u}} . \tag{A5.8}$$

In order to define the transformation matrices, it is obvious that kb and kc should be set to 1, and the remaining factors can be written as

$$k_a = -k_d = -\frac{L_{sp}}{L_{pp}} = k_i . \tag{A5.9}$$

The transformation matrices are finally defined as

$$\begin{pmatrix} \phi_i^u - \phi_i^l \\ \phi_i^s - \phi_i^l \end{pmatrix} = \begin{pmatrix} 1 & 0 \\ -k_i & 1 \end{pmatrix} \begin{pmatrix} V_i^{pp} \\ V_i^{tl} \end{pmatrix}, \quad \text{(A5.10a)}$$

$$\begin{pmatrix} i_i^u \\ i_i^s \end{pmatrix} = \begin{pmatrix} 1 & k_i \\ 0 & 1 \end{pmatrix} \begin{pmatrix} I_i^{pp} \\ I_i^{tl} \end{pmatrix}, \quad \text{(A5.10b)}$$

which correspond to Eqs. (4.30)-(4.31) in Section 4.6.1. The transformation is exact for homogeneous dielectric medium and ideal metallic conductors. For conductor loss and centered traces the transformation is still valid, however it is not exact for offset striplines in presence of lossy conductors, as well as for inhomogeneous dielectrics [45].

A.6. Formulation of the Via-Stripline Model

Figure 4.13(b) shows the equivalent circuit for 1-port via-to-stripline transition, derived from the modal decomposition procedure presented in Section 4.5. For the extended via and trace model, the ground terminals are made explicit, and the 6-terminal system is solved from the following relations (eliminating the index i for simplicity)

$$\phi^u - \phi^l = V^{pp}, \quad \text{(A6.1)}$$

$$\phi^s - \phi^l + k \cdot \left(\phi^u - \phi^l\right) = V^{tl}, \quad \text{(A6.2)}$$

$$I^u = k \cdot I^s + I^{pp}, \quad \text{(A6.3)}$$

$$-I^l = k \cdot I^s + I^{pp} + I^{tl}, \quad \text{(A6.4)}$$

$$\frac{I^{tl}}{V^{tl}} = Y^{tl}, \quad \text{(A6.5)}$$

$$\frac{I^{pp}}{V^{pp}} = Y^{pp} . \quad \text{(A6.6)}$$

Since $I^s = I^{tl}$, from Eqs. (A6.1-6), the system of equations is formulated for the currents I^u, I^l and I^s

$$\begin{aligned} I^u &= k \cdot I^{tl} + Y^{pp}\left(\phi^u - \phi^l\right) = k \cdot \left(Y^{tl} \cdot V^{tl}\right) + Y^{pp}\left(\phi^u - \phi^l\right) \\ &= k \cdot Y^{tl}\left(\phi^s - \phi^l + k \cdot \left(\phi^u - \phi^l\right)\right) + Y^{pp}\left(\phi^u - \phi^l\right) \\ &= k \cdot Y^{tl} \cdot \phi^s - k \cdot Y^{tl} \cdot \phi^l + k^2 \cdot Y^{tl} \cdot \phi^u - k^2 \cdot Y^{tl} \cdot \phi^l + Y^{pp} \cdot \phi^u - Y^{pp} \cdot \phi^l. \end{aligned}$$

$$I^u = \left(k^2 \cdot Y^{tl} + Y^{pp}\right) \cdot \phi^u + \left[\left(-k^2 - k\right) \cdot Y^{tl} - Y^{pp}\right] \cdot \phi^l + k \cdot Y^{tl} \cdot \phi^s. \quad \text{(A6.7)}$$

$$-I^{l} = k \cdot I^{s} + I^{pp} + I^{tl}$$
$$= \left(k+1\right) \cdot I^{tl} + I^{pp} = \left(k+1\right) \cdot Y^{tl} \cdot V^{tl} + Y^{pp} \cdot V^{pp},$$

$$I^{l} = -\left(k+1\right) \cdot Y^{tl} \cdot \left(\phi^{s} - \phi^{l} + k \cdot \left(\phi^{u} - \phi^{l}\right)\right) - Y^{pp} \cdot \left(\phi^{u} - \phi^{l}\right),$$

$$I^{l} = \left[\left(-k^{2} - k\right) \cdot Y^{tl} - Y^{pp}\right] \cdot \phi^{u} + \left[\left(k^{2} + 2k + 1\right) + Y^{pp}\right] \cdot \phi^{l} + \left[\left(-k-1\right) \cdot Y^{tl}\right] \cdot \phi^{s}. \quad \text{(A6.8)}$$

$$I^{s} = I^{tl} = Y^{tl} \cdot \left[\phi^{s} - \phi^{l} + k \cdot \left(\phi^{u} - \phi^{l}\right)\right],$$

$$I^{s} = I^{tl} = \left[k \cdot Y^{tl}\right] \cdot \phi^{u} + \left[\left(-k-1\right) \cdot Y^{tl}\right] \cdot \phi^{l} + \left[Y^{tl}\right] \cdot \phi^{s}. \quad \text{(A6.9)}$$

Equations (A6.7-A6.9) can be expressed as a Y-Matrix with the three terminal quantities explicitly defined [76]

$$\begin{pmatrix} I^{u} \\ I^{l} \\ I^{s} \end{pmatrix} = \begin{pmatrix} k^{2} \cdot Y^{tl} + Y^{pp} & \left(-k^{2} - k\right) \cdot Y^{tl} - Y^{pp} & k \cdot Y^{tl} \\ \left(-k^{2} - k\right) \cdot Y^{tl} - Y^{pp} & \left(k^{2} + 2k + 1\right) \cdot Y^{tl} + Y^{pp} & \left(-k-1\right) \cdot Y^{tl} \\ k \cdot Y^{tl} & \left(-k-1\right) \cdot Y^{tl} & Y^{tl} \end{pmatrix} \cdot \begin{pmatrix} \phi^{u} \\ \phi^{l} \\ \phi^{s} \end{pmatrix}. \quad \text{(A6.10)}$$

The expression in Eq. (A6.10) becomes Eq. (4.35) (for 1-port) if the signal terminal is defined as the reference potential. This is the case of interest for the models, according to the definition (Figure A.6)

$$I^{s} = I^{u} + I^{l}, \quad \text{(A6.11a)}$$

$$V^{us} = \phi^{u} - \phi^{s}, \quad \text{(A6.11b)}$$

$$V^{ls} = \phi^{l} - \phi^{s}. \quad \text{(A6.11c)}$$

With the definition in Eqs. (A6.11a to A6.11c), the system in Eq. (A6.10) can be reduced as follows

$$I^{u} = \left(k^{2} \cdot Y^{tl} + Y^{pp}\right) \cdot \left(V^{us} + \phi^{s}\right) + \left[\left(-k^{2} - k\right) \cdot Y^{tl} - Y^{pp}\right] \cdot \left(V^{ls} + \phi^{s}\right) + k \cdot Y^{tl} \phi^{s},$$

$$I^{u} = \left[k^{2} \cdot Y^{tl} + Y^{pp}\right] \cdot V^{us} + \left[\left(-k^{2} - k\right) \cdot Y^{tl} - Y^{pp}\right] \cdot V^{ls} + \underbrace{\left[k^{2} \cdot Y^{tl} + Y^{pp} + \left(-k^{2} - k\right) \cdot Y^{tl} - Y^{pp} + k \cdot Y^{tl}\right] \cdot \phi^{s}}_{=0}, \quad \text{(A6.12)}$$

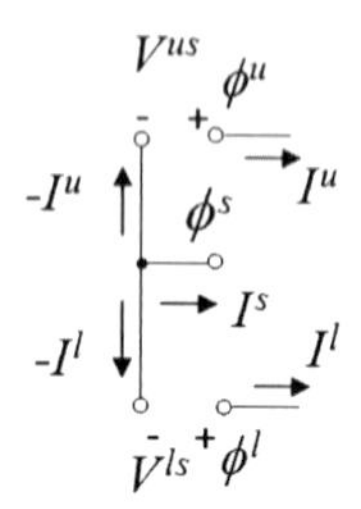

Figure A.6 Port definition for the via and trace model.

and

$$I^l = \left[\left(-k^2 - k\right) \cdot Y^{tl} - Y^{pp}\right] \cdot \left(V^{us} + \phi^s\right) + \left[\left(k^2 + 2k + 1\right) + Y^{pp}\right] \cdot \left(V^{gs} + \phi^s\right) + \left[\left(-k - 1\right) \cdot Y^{tl}\right] \cdot \phi^s,$$

$$I^l = \left[\left(-k^2 - k\right) \cdot Y^{tl} - Y^{pp}\right] \cdot V^{us} + \left[\left(k^2 + 2k + 1\right) + Y^{pp}\right] \cdot V^{ls} + \underbrace{\left[\left(-k^2 - k\right) \cdot Y^{tl} - Y^{pp} + \left(k^2 + 2k + 1\right) + Y^{pp} + \left(-k - 1\right) \cdot Y^{tl}\right] \cdot \phi^s}_{=0} . \tag{A6.13}$$

The admittance matrix is then reduced to

$$\begin{pmatrix} I^u \\ I^l \end{pmatrix} = \begin{pmatrix} k^2 \cdot Y^{tl} + Y^{pp} & (-k^2 - k) \cdot Y^{tl} - Y^{pp} \\ (-k^2 - k) \cdot Y^{tl} - Y^{pp} & (k^2 + 2k + 1) \cdot Y^{tl} + Y^{pp} \end{pmatrix} \cdot \begin{pmatrix} V^{us} \\ V^{ls} \end{pmatrix} . \tag{A6.14}$$

For N-ports, Y^{tl} and Y^{pp} become block matrices as shown in Eq. (4.35).

A.7. Segmentation Techniques

The segmentation technique [60] allows the concatenation of two network blocks with an arbitrary number of ports in terms of S- Z- *or* Y-parameters (Figure A.7). The ports need to be sorted, in a prior step, as extended or non-connected (p^a, p^b) and connected ports (q, r).

As depicted in Figure A.7, two network blocks A and B are connected by the ports q and r, respectively. The combined block C contains the non-connected ports p^a and p^b, whereas the connected ports are reduced. Each of these matrices can be expressed as

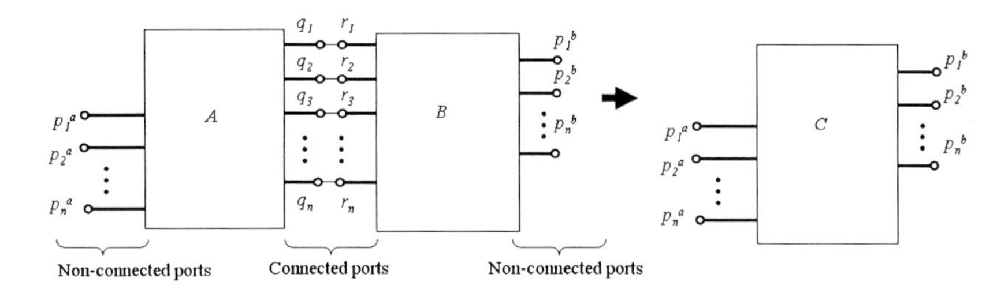

Figure A.7 Schematic representation of the segmentation method.

$$\overline{\overline{S}}_a = \begin{pmatrix} \overline{\overline{S}}_{p^a p^a} & \overline{\overline{S}}_{p^a q} \\ \overline{\overline{S}}_{q p^a} & \overline{\overline{S}}_{qq} \end{pmatrix}, \quad \overline{\overline{S}}_b = \begin{pmatrix} \overline{\overline{S}}_{p^b p^b} & \overline{\overline{S}}_{p^b r} \\ \overline{\overline{S}}_{r p^b} & \overline{\overline{S}}_{rr} \end{pmatrix}, \tag{A7.1}$$

$$\overline{\overline{S}}_{ab} = \overline{\overline{S}}_c = \begin{pmatrix} \overline{\overline{S}}_{p^a p^a} & \overline{\overline{S}}_{p^a p^b} \\ \overline{\overline{S}}_{p^b p^a} & \overline{\overline{S}}_{p^b p^b} \end{pmatrix}. \tag{A7.2}$$

It can be shown that the system of equations can be solved to find S_{AB} from S_a and S_b, according to [78]

$$\begin{aligned} \overline{\overline{S}}_{ab} = \overline{\overline{S}}_{pp} + \overline{\overline{S}}_{pq} \cdot \left(\overline{\overline{E}} - \overline{\overline{S}}_{rr} \overline{\overline{S}}_{qq} \right)^{-1} \left(\overline{\overline{S}}_{rp} + \overline{\overline{S}}_{rr} \overline{\overline{S}}_{qp} \right) + \\ \overline{\overline{S}}_{pr} \cdot \left(\overline{\overline{E}} - \overline{\overline{S}}_{qq} \overline{\overline{S}}_{rr} \right)^{-1} \left(\overline{\overline{S}}_{qp} + \overline{\overline{S}}_{qq} \overline{\overline{S}}_{rp} \right), \end{aligned} \tag{A7.3}$$

with

$$\overline{\overline{S}}_{pp} = \begin{pmatrix} \overline{\overline{S}}_{p^a p^a} & 0 \\ 0 & \overline{\overline{S}}_{p^b p^b} \end{pmatrix}, \quad \overline{\overline{S}}_{pq} = \begin{pmatrix} \overline{\overline{S}}_{p^a q} \\ 0 \end{pmatrix}, \quad \overline{\overline{S}}_{pr} = \begin{pmatrix} 0 \\ \overline{\overline{S}}_{p^b r} \end{pmatrix}, \tag{A7.3a}$$

$$\overline{\overline{S}}_{qp} = \left(\overline{\overline{S}}_{pq} \right)^T = \begin{pmatrix} \overline{\overline{S}}_{q p^a} & 0 \end{pmatrix}, \quad \overline{\overline{S}}_{rp} = \left(\overline{\overline{S}}_{pr} \right)^T = \begin{pmatrix} 0 & \overline{\overline{S}}_{r p^b} \end{pmatrix}, \tag{A7.3b}$$

and $\overline{\overline{E}}$ the identity matrix.

The size of the matrices and port order should be known before applying the segmentation procedure. The segmentation can also be carried out in terms of *Z*- or *Y*-parameters, for which [78]

$$\overline{\overline{Z}}_{ab} = \begin{pmatrix} \overline{\overline{Z}}_{p^a p^a} & 0 \\ 0 & \overline{\overline{Z}}_{p^b p^b} \end{pmatrix} + \begin{pmatrix} \overline{\overline{Z}}_{p^a q} \\ -\overline{\overline{Z}}_{p^b p^b} \end{pmatrix} \cdot \left(\overline{\overline{Z}}_{qq} + \overline{\overline{Z}}_{rr} \right)^{-1} \cdot \left(-\overline{\overline{Z}}_{qp^a} \quad \overline{\overline{Z}}_{rp^b} \right), \quad \text{(A7.4)}$$

$$\overline{\overline{Y}}_{ab} = \begin{pmatrix} \overline{\overline{Y}}_{p^a p^a} & 0 \\ 0 & \overline{\overline{Y}}_{p^b p^b} \end{pmatrix} - \left(\overline{\overline{Y}}_{pq} + \overline{\overline{Y}}_{pr} \right) \cdot \left(\overline{\overline{Y}}_{qq} + \overline{\overline{Y}}_{rr} \right)^{-1} \cdot \left(\overline{\overline{Y}}_{rp} \quad + \overline{\overline{Y}}_{qp} \right). \quad \text{(A7.5)}$$

The subscript definition in Eqs. (A7.4) and (A7.5) are analogous to the S-parameter case. The detailed mathematical derivation can be found in [78].

A.8. Mixed-Mode S-Parameters

The mixed-mode transformation is used to represent an S-parameter network in terms of its differential and common-mode parameters instead of the single-ended ones. A differential port is defined between two singled-ended ports as depicted in Figure A.8 for 4 ports.

The incident and reflected terms are defined for the differential (d) and common-mode (cm) cases as [133],[134]

$$a_{d(n-m)} = \frac{1}{\sqrt{2}}(a_n - a_m), \; b_{d(n-m)} = \frac{1}{\sqrt{2}}(b_n - b_m), \quad \text{(A8.1a)}$$

$$a_{cm(n-m)} = \frac{1}{\sqrt{2}}(a_n + a_m), \; b_{cm(n-m)} = \frac{1}{\sqrt{2}}(b_n + b_m), \quad \text{(A8.1b)}$$

The relations in Eq. (A8.1) can be expressed as

$$\begin{bmatrix} a_{d(n-m)} \\ a_{cm(n-m)} \end{bmatrix} = \overline{\overline{M}} \cdot \begin{bmatrix} a_n \\ a_m \end{bmatrix}, \quad \text{(A8.2a)}$$

$$\begin{bmatrix} b_{d(n-m)} \\ b_{cm(n-m)} \end{bmatrix} = \overline{\overline{M}} \cdot \begin{bmatrix} b_n \\ b_m \end{bmatrix}, \quad \text{(A8.2b)}$$

with the transformation matrix

$$\overline{\overline{M}} = \frac{1}{\sqrt{2}} \begin{bmatrix} 1 & -1 \\ 1 & 1 \end{bmatrix}. \quad \text{(A8.3)}$$

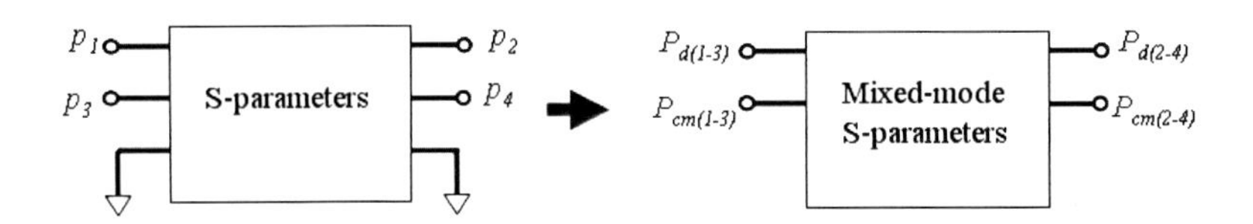

Figure A.8 Port definition for mixed-mode *S*-parameters, illustrated with a four port case, where *d* stands for differential and *cm* for common-mode.

For the 4-port case in Figure A.8, the transformation matrices become [134].

$$\begin{bmatrix} a_{d(1-3)} \\ a_{d(2-4)} \\ a_{cm(1-3)} \\ a_{cm(2-4)} \end{bmatrix} = \frac{1}{\sqrt{2}} \begin{bmatrix} 1 & 0 & -1 & 0 \\ 0 & 1 & 0 & -1 \\ 1 & 0 & 1 & 0 \\ 0 & 1 & 0 & 1 \end{bmatrix} \cdot \begin{bmatrix} a_1 \\ a_2 \\ a_3 \\ a_4 \end{bmatrix}, \tag{A8.4a}$$

$$\begin{bmatrix} b_{d(1-3)} \\ b_{d(2-4)} \\ b_{cm(1-3)} \\ b_{cm(2-4)} \end{bmatrix} = \frac{1}{\sqrt{2}} \begin{bmatrix} 1 & 0 & -1 & 0 \\ 0 & 1 & 0 & -1 \\ 1 & 0 & 1 & 0 \\ 0 & 1 & 0 & 1 \end{bmatrix} \cdot \begin{bmatrix} b_1 \\ b_2 \\ b_3 \\ b_4 \end{bmatrix}. \tag{A8.4b}$$

In general, the transformation between standard and mixed-mode (*mm*) *S*-parameters can be done in terms of the matrices $\overline{\overline{M}}$

$$\overline{\overline{S}}_{mm} = \overline{\overline{M}} \cdot \overline{\overline{S}} \cdot \overline{\overline{M}}^{-1}. \tag{A8.5}$$

The mixed-mode matrix is arranged by separating differential and common-modes, where the off-diagonal blocks represent the mode conversion between them (*d-cm* and *cm-d*).

$$\overline{\overline{S}}_{mm} = \begin{bmatrix} \overline{\overline{S}}_{d} & \overline{\overline{S}}_{d-cm} \\ \overline{\overline{S}}_{cm-d} & \overline{\overline{S}}_{cm} \end{bmatrix}. \tag{A8.6}$$

For the 4 port case, the following definition applies

$$\overline{\overline{S}}_{mixed\text{-mode}} = \begin{bmatrix} S_{d(1-3)d(1-3)} & S_{d(1-3)d(2-4)} & S_{d(1-3)cm(1-3)} & S_{d(1-3)cm(2-4)} \\ S_{d(2-4)d(1-3)} & S_{d(2-4)d(2-4)} & S_{d(2-4)cm(1-3)} & S_{d(2-4)cm(2-4)} \\ S_{cm(1-3)d(1-3)} & S_{cm(1-3)d(2-4)} & S_{cm(1-3)cm(1-3)} & S_{cm(1-3)cm(2-4)} \\ S_{cm(2-4)d(1-3)} & S_{cm(2-4)d(2-4)} & S_{cm(2-4)cm(1-3)} & S_{cm(2-4)cm(2-4)} \end{bmatrix}. \tag{A8.7}$$

If both single-ended (se) and differential ports exist, the following port arrangement can be used

$$\begin{bmatrix} \bar{b}_{d/cm} \\ \bar{b}_{se} \end{bmatrix} = \begin{bmatrix} \bar{\bar{S}}_{mm} & \bar{\bar{S}}_{mm-se} \\ \bar{\bar{S}}_{se-mm} & \bar{\bar{S}}_{se} \end{bmatrix} \cdot \begin{bmatrix} \bar{a}_{d/cm} \\ \bar{a}_{se} \end{bmatrix}, \tag{A8.8}$$

where the transformation matrix should be changed to [149]

$$\bar{\bar{M}}_c = \begin{bmatrix} \bar{\bar{M}} & 0 \\ 0 & \bar{\bar{E}} \end{bmatrix}, \tag{A8.9}$$

with $\bar{\bar{E}}$ the identity matrix. The expression to transform combined mixed-mode and singled-ended S-parameters can be written as

$$\bar{\bar{S}}_{mm/se} = \bar{\bar{M}}_c \cdot \bar{\bar{S}} \cdot \bar{\bar{M}}_c^{-1}, \tag{A8.10}$$

where

$$\bar{\bar{S}}_{mm/se} = \begin{bmatrix} \bar{\bar{S}}_{mm} & \bar{\bar{S}}_{mm-se} \\ \bar{\bar{S}}_{se-mm} & \bar{\bar{S}}_{se} \end{bmatrix}. \tag{A8.11}$$

A.9. The Contour Integral Method (CIM)

The CIM was originally developed for the analysis of microwave planar circuits [60]. It has been applied to calculate the impedance of the power buses of arbitrary shapes [61],[62],[152]. The merit of applying this method is to simplify the 3-D problem to a line integral, and hence reduce the numerical complexity.

Figure A.9 shows a pair of irregular power planes with C the boundary contour of the planes and C' the contour of the via barrels. r and r' are the observation and source points on both C and C'. $\hat{n}'$ and $\hat{t}'$ denote the unit normal and tangential vectors of both C and C'. Neglecting field variation in z-direction results in $\mathcal{E}_x = \mathcal{E}_y = \mathcal{H}_z = 0$. Therefore, the voltage between the planes can be defined as $V = -\mathcal{E}_z \cdot d$ with d as the substrate thickness. Together with the cylindrical wave solution and the scalar Green's theorem, a 2D contour integral equation can be derived [60]:

$$V(\mathbf{r}) = \frac{k}{2j} \cdot \oint_{C+C'} \left[\hat{\mathbf{R}} \cdot \hat{\mathbf{n}}' H_1^{(2)}\left(k|\mathbf{r}-\mathbf{r}'|\right) V(\mathbf{r}') - j\eta\, d\, H_0^{(2)}\left(k|\mathbf{r}-\mathbf{r}'|\right) \hat{\mathbf{n}}' \cdot \mathbf{J}(\mathbf{r}') \right] ds'. \tag{A9.1}$$

Where $J(r')$ is the current density on the contours. $H_0^{(2)}$ and $H_1^{(2)}$ are the *zeroth*-order and *first*-order Hankel functions of the second kind. $|r - r'|$ is the distance between source and observation points. $\hat{\mathbf{R}}$ represents the normalized vector of $r' - r$. k is the complex wave number including dielectric and ohmic losses. $\eta = \omega\mu_0 / k$ denotes the complex wave impedance with ω the angular frequency and μ_0 the free space permeability.

To apply the numerical procedure, the contour C is discretized into N segments with widths much smaller than the wavelength. Each of the segments can be considered as a line port. The boundary of each of the M vias in the contour C' is represented as a circular port. The voltages and currents can be assumed to be constant over the ports and thus pulse basis functions can be applied. The following $(N + M) \times (N + M)$ linear equation system is obtained:

$$\begin{bmatrix} \overline{\overline{U}}^{qq} & \overline{\overline{U}}^{qp} \\ \overline{\overline{U}}^{pq} & \overline{\overline{U}}^{pp} \end{bmatrix} \cdot \begin{bmatrix} \overline{V}^{q} \\ \overline{V}^{p} \end{bmatrix} = \begin{bmatrix} \overline{\overline{H}}^{qq} & \overline{\overline{H}}^{qp} \\ \overline{\overline{H}}^{pq} & \overline{\overline{H}}^{pp} \end{bmatrix} \cdot \begin{bmatrix} \overline{I}^{q} \\ \overline{I}^{p} \end{bmatrix}, \tag{A9.2}$$

where $\overline{V}^q$ and $\overline{V}^p$ are unknown voltage vectors of sizes $(N \times 1)$ and $(M \times 1)$ on the contours C and C', respectively, and $\overline{I}^q$, $\overline{I}^p$ are the excitation currents. The formulae to compute the sub-matrices in Eq. (A9.2) are available in [60],[62]. As a result of Eq. (A9.2), an impedance matrix can be obtained as

$$\begin{bmatrix} \overline{\overline{Z}}^{qq} & \overline{\overline{Z}}^{qp} \\ \overline{\overline{Z}}^{pq} & \overline{\overline{Z}}^{pp} \end{bmatrix} = \begin{bmatrix} \overline{\overline{U}}^{qq} & \overline{\overline{U}}^{qp} \\ \overline{\overline{U}}^{pq} & \overline{\overline{U}}^{pp} \end{bmatrix}^{-1} \cdot \begin{bmatrix} \overline{\overline{H}}^{qq} & \overline{\overline{H}}^{qp} \\ \overline{\overline{H}}^{pq} & \overline{\overline{H}}^{pp} \end{bmatrix}. \tag{A9.3}$$

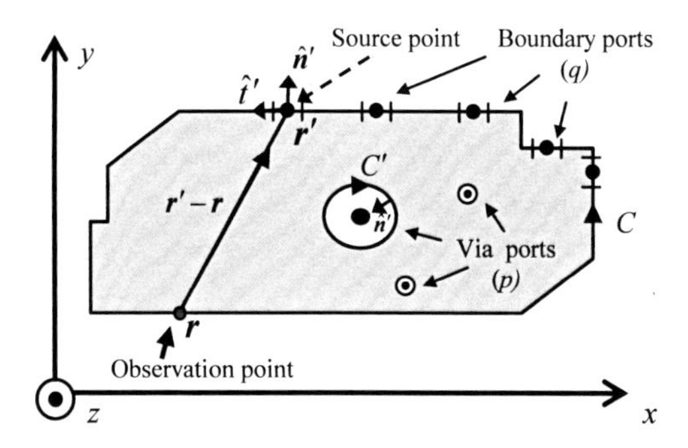

Figure A.9 Computation domain and variable definition for the contour integral method from [17].

Under the assumption of a perfect magnetic conductor (PMC) boundary condition at the plate edges, i.e., $\bar{I}^q = 0$, $\bar{\bar{Z}}^{pp}$ represents the parallel-plate impedance as defined in Section 4.3.1. The voltage distribution on the contour C can then be computed as

$$\bar{V}^q\Big|_{\bar{I}^q=0} = \bar{\bar{Z}}^{qp} \cdot \bar{I}^p. \tag{A9.4}$$

In addition, provided that the excitation currents are known, the voltage distribution inside the contour can be obtained as [61],[152]

$$V(\mathbf{r}) = \frac{k}{4j} \cdot \oint_{C+C'} \left[\hat{\mathbf{R}} \cdot \hat{\mathbf{n}}' H_1^{(2)}(k|\mathbf{r}-\mathbf{r}'|)\, V(\mathbf{r}') - j\eta d\, H_0^{(2)}(k|\mathbf{r}-\mathbf{r}'|)\, \hat{\mathbf{n}}' \cdot \mathbf{J}(\mathbf{r}')\right] ds'. \tag{A9.5}$$

A.10. The Equivalence Principle

According to the field equivalence principle [153], the radiated far fields from a pair of power planes can be approximated as the radiation from an equivalent magnetic surface current along the contour C [60]. The electric far field from the cavity is expressed as [153]:

$$\mathcal{E}_S(\mathbf{r}) \approx \frac{jk_0}{4\pi} \cdot \frac{e^{-jk_0 r}}{r} \sum_{i=1}^{N} V_q(\mathbf{r}_i') e^{jk_0 \mathbf{r}_i' \cdot \hat{\mathbf{e}}_r} \left(\hat{e}_r \times \hat{t}_i'\right) W_i, \tag{A10.1}$$

where k_0 is the free space wave number. $\hat{e}_r$ is the unit vector of r. $\mathbf{r}_i'$ is the location vector of the i-th port on the contour C. Voltages on boundary ports $V_q(\mathbf{r}_i')$ can be obtained by Eq. (A9.4).

To compute the radiation from a multilayer PCB, a Huygens equivalent surface can be applied that covers the whole PCB. The equivalent electric and magnetic surface currents on the outer sides of the top and bottom planes can usually be neglected [154]. Therefore, the radiated emission field from a multilayered PCB is approximated as that radiated by an array of parallel magnetic currents flowing around the sidewalls of the cavities. In a first approximation the edge coupling between cavities was neglected, thus the total electric far field generated from multilayered PCBs can be simply obtained by superposition of the fields radiated by the magnetic currents:

$$\mathcal{E}_M(\mathbf{r}) \approx \sum_{i=1}^{P} \mathcal{E}_S^i(\mathbf{r}), \tag{A10.2}$$

where P is the number of cavities, and $\mathcal{E}_S^i(\mathbf{r})$ denotes the radiated field from the i-th cavity, computed by Eq. (A10.1). To evaluate Eq. (A10.2), the voltage distribution on the boundary of each cavity has to be found, which was explained in the Section 6.3.1.

B. Code Architecture

The code developed as part of this work, called the Via Pin Field Simulation (VPF) Tool, is a prototype Matlab version that is based on the proposed models and the method discussed in Chapter 4. Figure B.1 illustrates the program functionality and the main code components. An interpreter reads the input files, which are a high-level description of the structure to be simulated. These files are then decoded. Another code component gets the variables created by the interpreter and identifies the cavities and their related interconnect elements. The calculator computes the parallel-plate impedance per cavity, generates or imports the transmission line models, and calculates or reads the via-to-plane capacitances. The calculator also combines the plane and trace models by applying modal decomposition, and creates the interconnection matrices for via-to-plane capacitances and lumped elements. Finally, the partial results are concatenated, for instance, using segmentation techniques. Post-processing functions are available to store and plot the results (e.g. as .sNp files [158]).

Figure B.2 shows the required input files. The *.board* file contains the element descriptions and port definition. It also links other files for stackup, component models, analysis options, etc.. The *.stk* (stack up) file contains the board profile definition, and the *.model* file the via description and options for the via-to-plane capacitance calculation. The *.matr* files define the material parameters, which can be frequency dependent. Analysis options are included in the *.apar* file.

More details of the code implementation, definition of input files and examples are documented in [159].

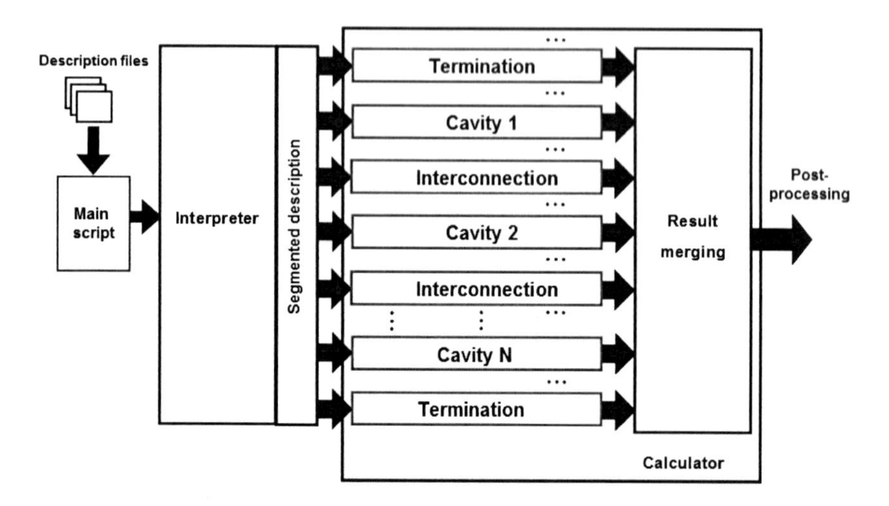

Figure B.1 General description of the via pin field simulation tool (VPF) and its main code components.

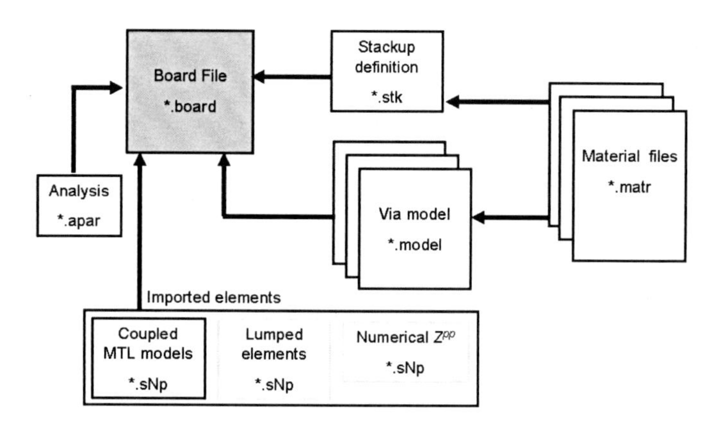

Figure B.2 Required input files for the high-level description of a multilayer substrate.

C. Additional Results

C.1. Results up to 40 GHz for the Examples in Section 5.2.6

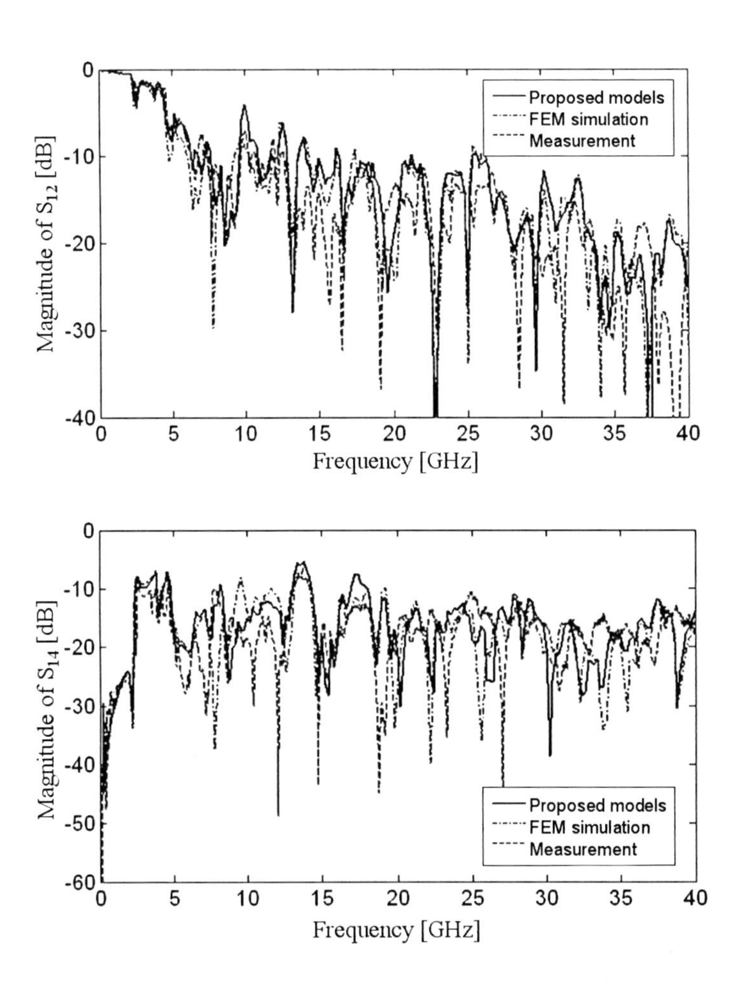

Figure C.1 Magnitude of *S*-parameters for the TV-1 in Figure 5.15, obtained by measurement, FEM full-wave simulation and the models up to 40 GHz.

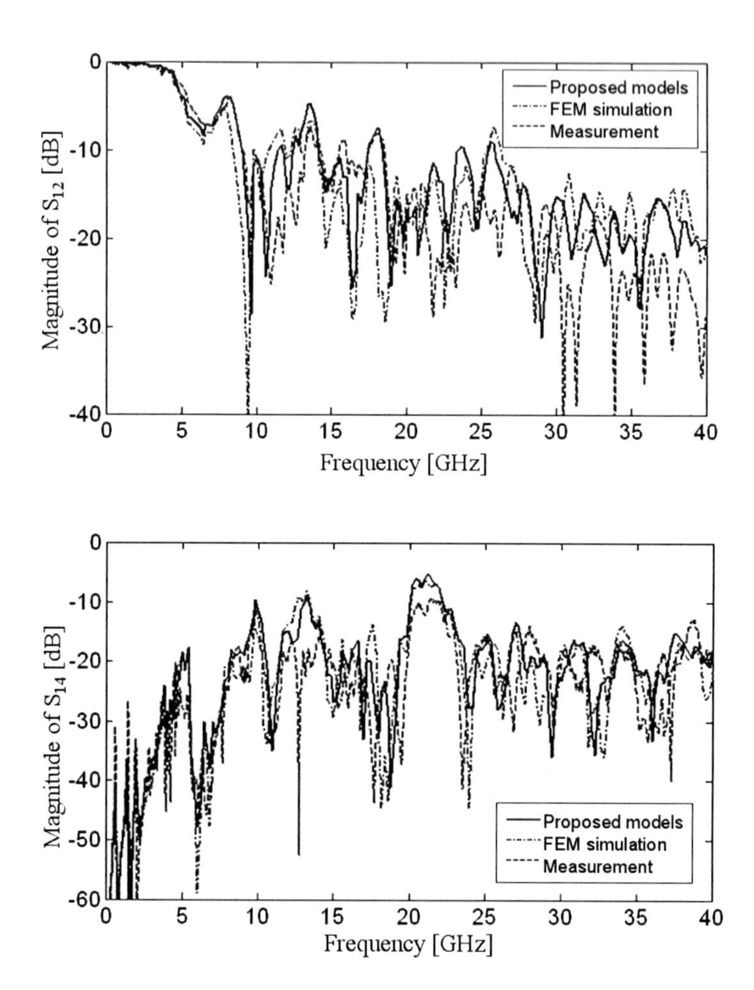

Figure C.2 Magnitude of *S*-parameters for the TV-2 in Figure 5.15, obtained by measurement, FEM full-wave simulation and the models up to 40 GHz.

C.2. Results up to 40 GHz for the Example in Section 5.3.2

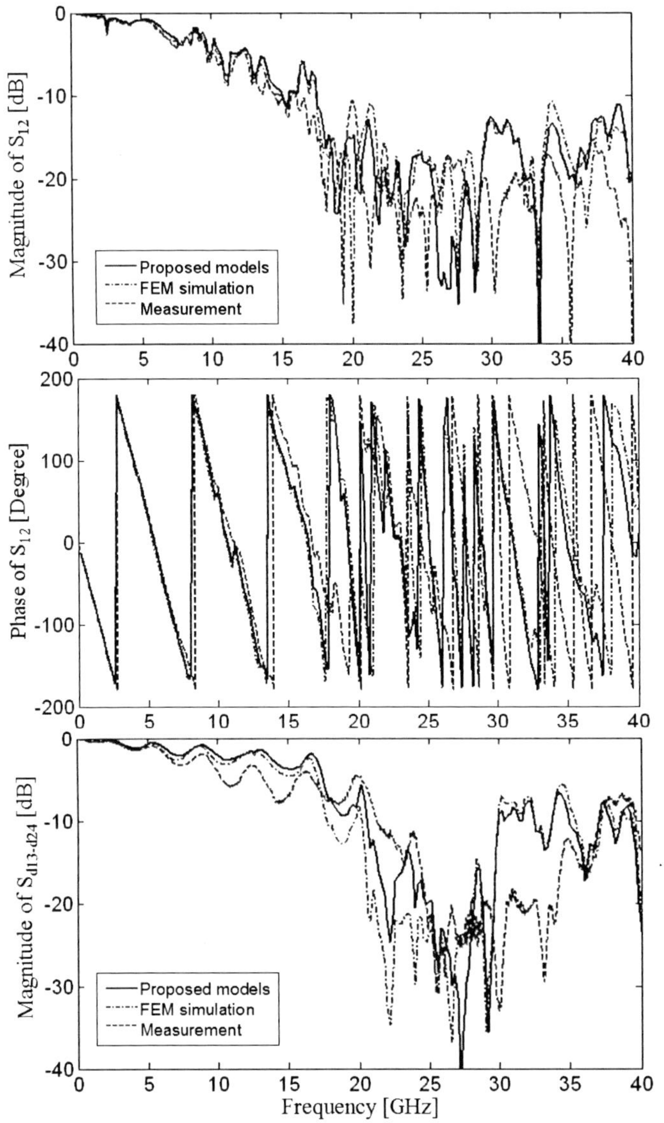

Figure C.3 *S*-parameter single-ended and differential transmission for the TV-3 in Figure 5.22, obtained by measurement, FEM full-wave simulation and the models, up to 40 GHz.

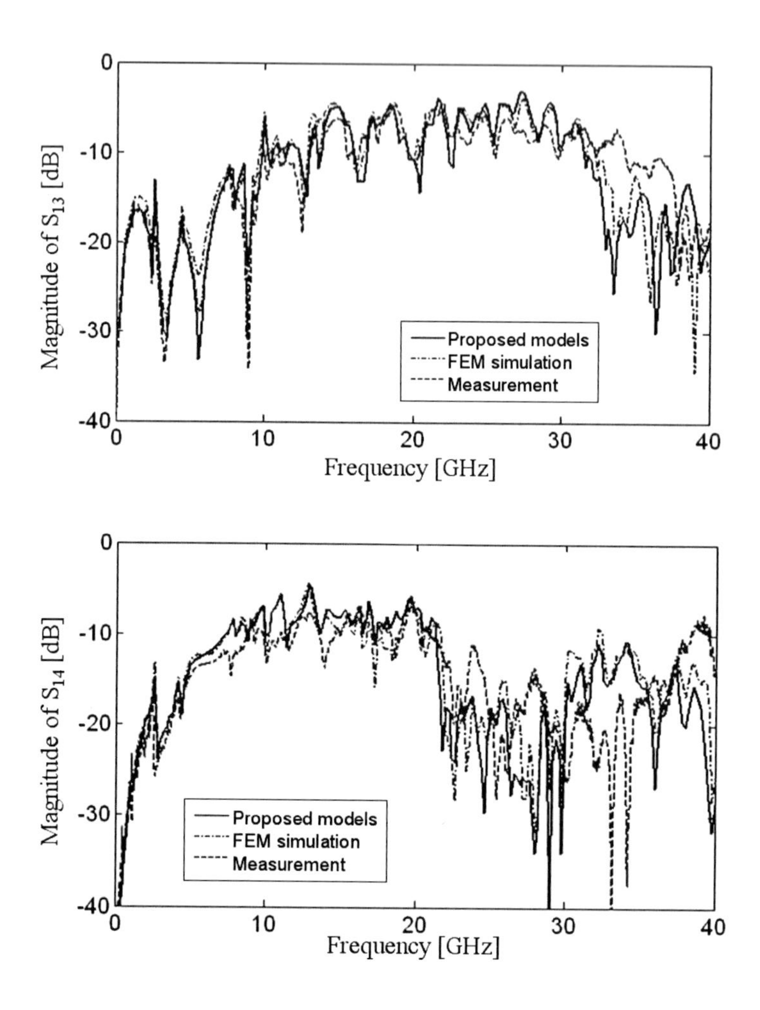

Figure C.4 Single-ended crosstalk parameters for the TV-3 in Figure 5.22, obtained by measurement, FEM full-wave simulation and the models, up to 40 GHz.

References

[1] H. Harrer, D. M. Dreps, T.-M Winkel, W. Scholz, B. G. Truong, A. Huber, T. Zhou, K. L. Christian, and G. F. Goth, "High-speed interconnect and packaging design of the IBM System z9 processor cage," *IBM Journal of Research and Development*, vol. 51, no. 1/2, pp. 37-52, January/March 2007.

[2] P. Coteus, H. R. Bickford, T. M. Cipolla, P. G. Crumley, A. Gara, S. A. Hall, G. V. Kopcsay, A. P. Lanzetta, L. S. Mok, R. Rand, R. Swetz, T. Takken, P. La Rocca, C. Marroquin, P. R. Germann, and M. J. Jeanson, "Packaging of the Blue Gene/L system architecture," *IBM Journal of Research and Development*, vol. 49, no. 2/3, pp. 213-248, March/May 2005.

[3] D. G. Kam, M. B. Ritter, T. J. Beukema, J. F. Bulzacchelli, P. K. Pepeljugoski, Y. H. Kwark, L. Shan, X. Gu, C. W. Baks, R. A. John, G. Hougham, C. Schuster, R. Rimolo-Donadio, and B. Wu, "Is 25 Gb/s on-board signaling viable?," *IEEE Transactions on Advanced Packaging*, vol. 32, no. 2, pp. 328-344, May 2009.

[4] D. Derickson and M. Müller, "Digital communications test and measurement: high-speed physical layer characterization," Upper Saddle River, NJ, USA: *Prentice Hall*, 2007.

[5] V. Stojanovic and M. Horowitz, "Modeling and analysis of high-speed links," *in Proc. IEEE Custom Integrated Circuits Conference*, pp. 589-594, San Jose, California, USA, September 21-24, 2003.

[6] International Technology Roadmap for Semiconductors, *Executive summary 2009 edition*, pp. 15-26. 2009 [Online]. Information available: http://www.itrs.net (March 2010).

[7] R. R. Tummala, "Fundamentals of microsystem packaging," New York, USA: *McGraw-Hill Professional*, 2004.

[8] S. Müller, R. Rimolo-Donadio, M. Kotzev, H.-D. Brüns, and C. Schuster, "Effect of mixed-reference planes on single-ended and different links in multilayer substrates," *in Proc. 14th IEEE Workshop on Signal Propagation on Interconnects SPI*, Hildesheim, Germany, May 10-12, 2010.

[9] X. Duan, R. Rimolo-Donadio, H.-D. Brüns, and C. Schuster, "Fast and concurrent simulations for SI, PI, and EMI analysis of multilayer printed circuit boards," *in Proc. Asia-Pacific Symposium on Electromagnetic Compatibility*, invited paper, Beijing, China, April 12-16, 2010.

[10] S. Müller, R. Rimolo-Donadio, H.-D. Brüns, and C. Schuster, "Schnelle Simulation verlustbehafteter Verbindungsstrukturen auf Leiterplatten auf der Grundlage quasianalytischer Via-Modelle und der Leitungstheorie," *in Proc. Internationale Fachmesse und Kongress für Elektromagnetische Verträglichkeit EMV*, Düsseldorf, Germany, March 9-11, 2010.

[11] X. Gu, R. Rimolo-Donadio, Z. Yu, F. de Paulis, Y. H. Kwark, M. Cocchini, M.B. Ritter, B. Archambeault, A. Ruehli, J. Fan, and C. Schuster, "Fast-physics-based via and trace models for signal and power integrity co-analysis," *in Proc. IEC DesignCon Conference*, Santa Clara, USA, February 1-4, 2010. [Best Paper Award].

[12] R. Rimolo-Donadio, X. Duan, H.-D. Brüns, and C. Schuster, "Comprehensive multilayer substrate models for co-simulation of power and signal integrity," *in Proc. 42th International Symposium on Microelectronics IMAPS*, San Jose, California, USA, November 1-5, 2009.

[13] R. Rimolo-Donadio, X. Duan, H.-D. Brüns, and C. Schuster, "Differential to common mode conversion due to asymmetric ground via configurations," *in Proc. 13th IEEE Workshop on Signal Propagation on Interconnects SPI, Strasbourg*, France, May 12-15, 2009, pp. 1-4.

[14] X. Gu, F. De Paulis, R. Rimolo-Donadio, K. Shringarpure, Y. Zhang, B. Archambeault, S. Connor, Y. H. Kwark, M. B. Ritter, J. Fan, and C. Schuster, "Fully analytical methodology for fast end-to-end link analysis on complex printed circuit boards including signal and power integrity effects," *in Proc. IEC DesignCon Conference*, Santa Clara, USA, February 2-5, 2009.

[15] R. Rimolo-Donadio, H.-D. Brüns, and C. Schuster, "Including stripline connections into network parameter based via models for fast simulation of interconnects," *in Proc. 20th International Zurich Symposium on Electromagnetic Compatibility*, Zurich, Switzerland, January 12-16, 2009, pp. 345-348.

[16] R. Rimolo-Donadio, A. J. Stepan, H.-D. Brüns, J. L. Drewniak, and C. Schuster, "Simulation of via interconnects using physics-based models and microwave network parameters," *in Proc. 12th IEEE Workshop on Signal Propagation on Interconnects SPI*, Avignon, France, May 12-15, 2008.

[17] X. Duan, R. Rimolo-Donadio, H.-D. Brüns, and C. Schuster, "A combined method for fast analysis of signal propagation, ground noise, and radiated emission of multilayer printed circuit boards," *IEEE Transactions on Electromagnetic Compatibility*, vol. 52, no. 2, pp. 487-495, May 2010.

[18] R. Rimolo-Donadio, H.-D. Brüns, and C. Schuster, "Hybrid approach for efficient calculation of the parallel-plate impedance of lossy power/ground planes," *Microwave and Optical Technology Letters*, vol. 51, no. 9, pp. 2051-2056, September 2009. [See also [28]]

[19] R. Rimolo-Donadio, X. Gu, Y. H. Kwark, M. B. Ritter, B. Archambeault, F. D. Paulis, Y. Zhang, J. Fan, H.-D. Brüns, and C. Schuster, "Physics-based via and trace models for efficient link simulation on multilayer structures up to 40 GHz," *IEEE Transactions on Microwave Theory and Techniques*, vol. 57, no. 8, pp. 2072-2083, August 2009.

[20] M. Kotzev, R. Rimolo-Donadio, H.-D. Brüns, and C. Schuster, "Multiport measurement and deembedding techniques for crosstalk study in via arrays," *in Proc. 14th IEEE Workshop on Signal Propagation on Interconnects SPI*, Hildesheim, Germany May 10-12, 2010.

[21] X. Duan, R. Rimolo-Donadio, H.-D. Brüns, B. Archambeault, and C. Schuster, "Special session on power integrity techniques: contour integral method for rapid computation of power/ground plane impedance," *in Proc. IEC DesignCon Conference*, Santa Clara, USA, February 1-4, 2010.

[22] M. Kotzev, R. Rimolo-Donadio, and C. Schuster, "Extraction of broadband error boxes for microprobes and recessed probe launches for measurement of printed circuit board structures," *in Proc. 13th IEEE Workshop on Signal Propagation on Interconnects SPI*, Strasbourg, France, May 12-15, 2009, pp. 1-4.

[23] Y. Zhang, R. Rimolo-Donadio, C. Schuster, E. Li, and J. Fan, "Extraction of via-plate capacitance of an eccentric via by an integral approximation method," *IEEE Microwave and Wireless Components Letters*, vol. 19, no. 5, pp. 275-277, May 2009.

[24] M. Kotzev, X. Gu, Y. H. Kwark, M. B. Ritter, R. Rimolo-Donadio, and C. Schuster, "Bandwidth study of recessed probe launch variations for broadband measurement of embedded PCB structures," *in Proc. IEEE German Microwave Conference GeMic*, Munich, Germany, March 16-18, 2009, pp. 1-4.

[25] M. B. Ritter, P. Pepeljugoski, X. Gu, Y. Kwark, D. Kam, R. Rimolo-Donadio, B. Wu, C. Baks, R. John, L. Shan, and C. Schuster, "The viability of 25 Gb/s on-board signaling," *in Proc. IEEE Electronic Components and Technology Conference ECTC*, Lake Buena Vista, Florida, USA, May 27-30, 2008, pp. 1121-1127.

[26] R. Rimolo-Donadio, C. Schuster, Y. Kwark, X. Gu, and M. B. Ritter, "Analysis and optimization of the recessed probe launch for high frequency measurements of PCB interconnects," *in Proc. IEEE Design, Automation and Test in Europe Conference DATE*, Munich, Germany, March 10-14, 2008, pp. 252-255. [Best Interactive Presentation Award]

[27] V. Vahrenholt, H.-D. Brüns, H. Singer, and R. Rimolo-Donadio, "Verkopplung einer schnellen PEEC-Methode mit der Momentenmethode bei gedruckten Schaltungen mittels der elektrischen Feldstärke," *in Proc. Internationale Fachmesse und Kongress für Elektromagnetische Verträglichkeit EMV*, Düsseldorf, Germany, February 18-21, 2008.

[28] R. Rimolo-Donadio, H.-D. Brüns, and C. Schuster, "Erratum: Hybrid approach for efficient calculation of the parallel-plate impedance of lossy power/ground planes," *Microwave and Optical Technology Letters*, vol. 52, no. 1, pp. 247, January 2010.

[29] W. J. Dally and J. W. Poulton, "Digital Systems Engineering," Cambridge, United Kingdom: *Cambridge University Press*, 1998.

[30] W. M. Green, M. J. Rooks, L. Sekaric, and Y. A. Vlasov, "Ultra-compact, low RF power, 10 Gb/s silicon Mach-Zehnder modulator," *Optics Express*, vol. 15, issue 25, pp. 17106-17113, December 2007.

[31] A. F. Brenner, M. Ignatowski, J. A. Kash, D. M. Kuchta, and M. B. Ritter, "Exploitation of optical interconnects in future server architectures," *IBM Journal of Research and Development*, vol. 49, no. 4/5, pp. 755-755, July/September 2005.

[32] R. R. Tummala and M. Swaminathan, "Introduction to system-on-package (SOP). Miniaturization of the entire system," New York, USA: *McGraw-Hill*, 2008.

[33] U. Knickerbocker, P. S. Andry, L. P. Buchwalter, A. Deutsch, R. R. Horton, K. A. Jenkins, Y. H. Kwark, G. McVicker, C. S. Patel, R. J. Polastre, C. Schuster, A. Sharma, S. M. Sri-Jayantha, C. W. Surovic, C. K. Tsang, B. C. Webb, S. L. Wright, S. R. McKnight, E. J. Sprogis, and B. Dang, "Development of the next-generation system-on-package (SOP) technology based on silicon carriers with fine-pitch chip interconnection," *IBM Journal of Research and Development*, vol. 49, no. 4/5, pp. 725-754, July/September, 2005.

[34] M. Popovich, A. V. Mezhiba, and E. G. Friedman, "Power distribution networks with on-chip decoupling capacitors," New York, USA: *Springer*, 2008.

[35] E. D. Blackshear, M. Cases, E. Klink, S. R. Engle, R. S. Malfatt, D. N. de Araujo, S. Oggioni, L. D. LaCroix, J. K. Wakil, G. G. Hougham, N. H. Pham, and D. J. Russell, "The evolution of build-up package technology and its design challenges," *IBM Journal of Research and Development*, vol. 49, no. 4/5, pp. 641-661, July/September, 2005.

[36] R. Chanchani, [D. Lu, C. P. Wong, Ed.], "Materials for advanced packaging. Chapter 1: 3D integration technologies – An overview," New York, USA: *Springer*, 2008.

[37] C. F. Coombs, Ed., "Printed circuits handbook," 6th Edition, New York, USA: *McGraw-Hill*, 2008.

[38] C. A. Harper [Ed.], "High performance printed circuits boards," New York, USA: *McGraw-Hill Professional Engineering*, 2000.

[39] B. Young, "Digital signal integrity: modeling and simulation with interconnect and packages," London, U.K.: *Prentice Hall Modern Semiconductors Design Series*, 2000.

[40] J. Fan, X. Ye, J. Kim, B. Archambeault, and A. Orlandi, "Signal integrity design for high-speed digital circuits: progress and directions," *IEEE Transactions on Electromagnetic Compatibility*, vol. 52, issue 2, pp. 392-400, May 2010.

[41] H. W. Johnson and M. Graham, "High-Speed Digital Design: A handbook of black magic", New Jersey, USA: *Prentice Hall*, 1993.

[42] P. Noel, F. Zarkeshvari, and T. Kwasniewski, "Recent advances in high-speed serial I/O trends, standards and techniques," *in Proc. Canadian Conference on Electrical and Computer Engineering*, Saskatoon, Saskatchewan, Canada, May 1-4, 2005, pp. 1292-1295.

[43] J. F. Bulzacchelli, M. Meghelli, S. V. Rylov, W. Rhee, A. V. Rylyakov, H. A. Ainspan, B. D. Parker, M. P. Beakes, A. Chung, T. J. Beukema, P. K. Pepeljugoski, L. Shan, Y. H. Kwark, S. Gowda, and D. J. Friedman, "A 10Gb/s 5-Tap DFE/ 4-Tap FFE Transceiver in 90nm CMOS Technology," *IEEE Journal of Solid-State Circuits*, vol. 41, no. 12, pp. 2885-2900, December 2006.

[44] T. Beukema, M. Sorna, K. Selander, S. Zier, B. J. Ji, P. Murfet, J. Mason, W. Rhee, H. Ainspan, B. Parker, and M. Beakes, "A 6.4-Gb/s CMOS SerDes core with feed-forward and decision-feedback equalization," *IEEE Journal of Solid-State Circuits*, vol. 40, no. 12, pp. 2633-2645, December 2005.

[45] M. Swaminathan and A. Ege Engin, "Power integrity modeling and design for semiconductors and systems," Upper Saddle River, NJ, USA: *Prentice Hall*, 2008.

[46] L. W. Schaper, S. Ang, Y. L. Low, and D. R. Oldham, "Electrical characterization of the interconnected mesh power system (IMPS) MCM topology," *IEEE Transactions on Components, Packaging and Manufacturing Technology -part B*, vol.18, no. 1, pp. 99-105, February 1995.

[47] P. Larsson, "di/dt noise in CMOS integrated circuits," *Analog Integrated Circuits and Signal Processing*, vol. 14, pp. 113-129, September 1997.

[48] J. L. Knighten, B. Archambeault, J. Fan, G. Selli, S. Connor, and J. L. Drewniak, "PDN Design Strategies: I. Ceramic SMT decoupling capacitors – what values should I choose?," *IEEE EMC Society Newsletter*, issue no. 207, pp. 54-64, Fall 2005.

[49] T. H. Hubing, J. L. Drewniak, T. P. Van Doren, and D. M. Hockanson, "Power bus decoupling on multilayer printed circuit boards," *IEEE Transactions on Electromagnetic Compatibility*, vol. 37, no. 2, pp. 155-166, May 1995.

[50] C. R. Paul, "Introduction to electromagnetic compatibility," 2 Ed. New York, USA: *Wiley-Interscience*, 2006.

[51] S. Caniggia and F. Maradei, "Signal integrity and radiated emission of high-speed digital systems," United Kingdom: *John Wiley & Sons*, 2008.

[52] H. Johnson and M. Graham, "High-speed signal propagation: advanced black magic", New Jersey, USA: *Prentice Hall*, 2003.

[53] B. C. Wadell "Transmission line design handbook", Norwood, MA, USA: *Artech House*, 1991.

[54] C. Schuster and W. Fichtner, "Parasitic modes on printed circuit boards and their effects on EMC and signal integrity," *IEEE Transactions Electromagnetic Compatibility*, vol. 43, no. 4, pp. 416-425, November 2000.

[55] B. Archambeault and A. E. Ruehli, "Analysis of power/ground-plane EMI decoupling performance using the partial-element equivalent circuit technique," *IEEE Transactions Electromagnetic Compatibility*, vol. 43, no. 4, pp. 437-445, November 2001.

[56] J. Yook, N. I. Dibb, and L. P. B. Katehi, "Characterization of high frequency interconnects using the finite difference time domain and finite element methods", *IEEE Transactions on Microwave Theory and Techniques*, vol. 42, no. 9, pp. 1727-1735, September 1994.

[57] T. Okoshi and T. Miyoshi, "The planar circuit – An approach to microwave integrated circuitry," *IEEE Transactions on Microwave Theory and Techniques*, vol. 20, no. 4, pp. 245-252, April 1972.

[58] R. Sorrentino, "Planar circuits, waveguide methods, and segmentation method," *IEEE Transactions on Microwave Theory and Techniques*, vol. 33, no. 10, pp. 1057-1066, October 1985.

[59] A. Ege Engin, K. Bharath, and M. Swaminathan, "Multilayered finite-difference method (MFDM) for modeling of package and printed circuit board planes," *IEEE Transactions on Electromagnetic Compatibility*, vol. 49, no. 2, pp. 441-447, May 2007.

[60] T. Okoshi, "Planar Circuits for Microwaves and Lightwaves", Berlin, Germany: *Springer-Verlag*, 1985.

[61] M. Stumpf and M. Leone, "Efficient 2-D integral equation approach for the analysis of power bus structures with arbitrary shape," *IEEE Transactions Electromagnetic on Compatibility*, vol. 51, no. 1, pp. 38-45, February 2009.

[62] X. Wei, E. Li, E. Liu, and X. Cui, "Efficient modeling of rerouted return currents in multilayered power-ground planes by using integral equation," *IEEE Transactions on Electromagnetic Compatibility*, vol. 50, No. 3, pp.740-743, August 2008.

[63] J. H. Kim and M. Swaminathan, "Modeling of irregular shaped power distribution planes using transmission matrix method," *IEEE Transactions on Advanced Packaging*, vol. 25, no. 2, pp. 189-199, May 2002.

[64] J. Park, H. Kim, Y. Jeong, J. Kim, J. S. Pak, D. G. Kam, and J. Kim, "Modeling and measurement of simultaneous switching noise coupling through signal via transition," *IEEE Transactions on Advanced Packaging*, vol. 29, no. 3, pp. 548-559, August 2006.

[65] L. Tsang, H. Chen, C. C. Huang, and V. Jandhyala, "Modeling of multiple scattering among vias in planar waveguides using foldy-lax equations," *Microwave Optical Technology Letters*, vol. 31, no. 4, pp. 375-384, Nov. 2004.

[66] H. Chen, Q. Li, L. Tsang, C. C. Huang, and V. Jandhyala, "Analysis of a large number of vias and differential signaling in multilayered structures," *IEEE Transactions on Microwave Theory and Techniques*, vol. 51, no. 3, pp. 818-829, March 2003.

[67] Z. Z. Oo, E. X. Liu, E. P. Li, X. Wei, Y. Zhang, M. Tan, L. W. J. Li, and R. Vahldieck, "A semi-analytical approach for system-level electrical modeling of electronic packages with large number of vias," *IEEE Transactions on Advanced Packaging*, vol. 31, no. 2, pp. 267-274, May 2008.

[68] E. X. Liu, E. P. Li, Z. Z. Oo, X. C. Wei, Y. J. Zhang, and R. Vahldieck, "Novel methods for modelling of multiple vias in multilayered parallel-plate structures," *IEEE Transactions on Microwave Theory and Techniques*, vol. 57, no. 7, pp. 1724-1733, July 2009.

[69] G. T. Lei, R. W. Techentin, P. R. Hayes, D. J. Schwab, and B. K. Gilbert, "Wave model solution to the ground/power plane noise problem," *IEEE Transactions on Instrumentation and Measurement*, vol. 44, no. 2, pp. 300-303, April 1995.

[70] R. Ito, R. W. Jackson, and T. Hongsmatip, "Modelling of interconnections and isolation within a multilayered ball grid array package," *IEEE Transactions on Microwave Theory and Techniques*, vol. 47, no. 9, pp. 1819-1825, September 1999.

[71] G. Selli, C. Schuster, Y. Kwark, M. Ritter, and J. L. Drewniak, "Model-to-hardware correlation of physics based via models with the parallel plate impedance included," *in Proc. IEEE Symposium Electromagnetic Compatibility*, Portland, Oregon, USA, August 2006, pp. 781-785.

[72] Z. Z. Oo, E. X. Liu, E. P. Li, X. Wei, E. X. Liu, Y. J. Zhang, and L. W. J. Li, "Hybridization of the scattering matrix method and modal decomposition for analysis of signal traces in a power distribution network," *IEEE Transactions on Electromagnetic Compatibility*, vol. 51, no. 3, pp. 784-791, August 2009.

[73] X. C. Wei, E. P. Li, E. X. Liu, and R. Vahldieck, "Efficient simulation of power distribution network by using integral-equation and modal-decoupling technology," *IEEE Transactions on Microwave Theory and Techniques*, vol. 56, no. 10, pp. 2277- 2285, October 2008.

[74] J. Kim, Y. Joeng, J. Kim, J. Lee, C. Ryu, J. Shim, M. Shin, and J. Kim, "Modeling and measurement of interlevel electromagnetic coupling and fringing effect in a hierarchical power distribution network using segmentation method with resonant cavity model," *IEEE Transactions on Advanced Packaging*, vol. 31, no. 3, pp. 544–557, Aug. 2008.

[75] X. C. Wei, E. P. Li, E. X. Liu, E. K. Chua, Z. Z. Oo, and R. Vahldieck, "Emission and susceptibility modeling of finite-size power-ground planes using a hybrid integral equation method" *IEEE Transactions on Advanced Packaging*, vol. 31, no. 3, pp. 536-543, August 2008.

[76] A. E. Engin, K. Bharath, M. Swaminathan, M. Cases, B. Mutnury, N. Pham, D. N. de Araujo, and E. Matoglu, "Finite-difference modeling of noise coupling between power/ground planes in multilayered packages and boards," *in Proc. 56th Electronic Components Technology Conference ECTC*, San Diego, California, USA, May 30-June 2, 2006, pp. 1262–1267.

[77] Y. Chen, Z. Chen, Y. Wu, D. Xue, and J. Fang, "A new approach to signal integrity analysis of high-speed packaging," *in Proc. IEEE 4th Topical Meeting on Electrical Performance of Electronic Packaging EPEP*, Portland, USA, October 2-4, 1995, pp. 235-238.

[78] K. C. Gupta and M. D. Abouzahra, Ed., "Analysis and Design of Planar Microwave Components," New Jersey, USA: *IEEE Press*, Chapter 3, pp. 75-86, 2004.

[79] A. Ege Engin, W. John, G. Sommer, W. Mathis, and H. Reichl, "Modeling of striplines between a power and a ground plane," *IEEE Transactions on Advanced Packaging*, vol. 29, no. 3, pp. 415–426, August 2006.

[80] C. Schuster, Y. H. Kwark, G. Selli, and P. Muthana, "Developing a "physical" model for vias", *in Proc. IEC DesignCon 2006*, Santa Clara, California, USA, February 6-9, 2006.

[81] L. Shan, Y. H. Kwark, D. Dreps, and J. Trewhella, "How detrimental could a via be?," *in Proc. IEEE 13th Topical Meeting on Electrical Performance of Electronic Package EPEP*, Portland, USA, October 25-27, 2004, pp. 91-91.

[82] T. Wang, R. F. Harrington, and J .R. Mautz, "Quasi-static analysis of a microstrip via through a hole in a ground plane," *IEEE Transactions on Microwave Theory and Techniques*, vol. 36, no. 6, pp. 1008-1013, June 1988.

[83] CST Corporation, CST Microwave Studio (MWS) Ver. 2008/2009 [Online], Darmstadt, Germany. Information available: http://www.cst.com (March 2010).

[84] C. Schuster, "Effects of parallel planes," *Lecture notes on electrical design and characterisation of packages and interconnects*, Technische Universitaet Hamburg-Harburg, 2010.

[85] S. Ramo, J. R. Whinnery, and T. Van Duzer, "Fields and waves in communications electronics," 3rd Ed., New York, USA: *John Wiley & Sons*, 1993.

[86] C. A. Balanis, "Advanced engineering electromagnetics," New York, USA: *John Wiley and Sons*, 1989.

[87] S. Wu, X. Chang, C. Schuster, X. Gu, and J. Fan, "Eliminating via-plane coupling using ground vias for high-speed signal transitions," *in Proc. IEEE Electrical Performance of Electronic Package Conference EPEP*, San Jose, California, USA, October 27-29, 2008, pp. 247-250.

[88] E. R. Pillai, "Coax via— A technique to reduce crosstalk and enhance impedance match at vias in high-frequency multilayer packages verified by FDTD and MoM modeling," *IEEE Transactions on Microwave Theory and Techniques*, vol. 45, no. 10, pp. 1981-1985, October 1997.

[89] X. Gu, A. E. Ruehli, and M. B. Ritter, "Impedance design for multi-layered vias," *in Proc. IEEE Electrical Performance of Electronic Package Conference EPEP*, San Jose, California, USA, October 27-29, 2008, pp. 317-320.

[90] W. R. Eisenstadt, and Y. Eo, "S-parameter-based IC interconnect transmission line characterization," *IEEE Transactions on Components, Hybrids, and Manufacturing Technology*, vol. 15, no. 4, pp. 483-490, August 1992.

[91] T. Neu, "Designing controlled-impedance vias," *in Electronics Design, Strategy, News EDN Magazine*, pp. 67-72, October 2003.

[92] T. Kushta, K. Narita, T. Kaneko, T. Saeki, and H. Tohya, "Resonance stub effect in a transition from a through via hole to a stripline in multilayer PCBs," *IEEE Microwave Wireless Component Letters*, vol. 13, no. 5, pp. 169-171, May 2003.

[93] Q. Gu, Y. E. Yang, and M. A. Tassoudji, "Modelling and analysis of vias in multilayered integrated circuits," *IEEE Transactions on Microwave Theory and Techniques*, vol. 41, no. 2, pp. 206-214, February 1993.

[94] C. Schuster, G. Selli, Y. H. Kwark, M. B. Ritter, J. L. Drewniak, "Progress in representation and validation of physics-based via models", *in Proc. 11th IEEE Workshop Signal Propagation on Interconnects*, Genova, Italy, May 13-16, 2007, pp. 145-148.

[95] C. Schuster, G. Selli, Y. H. Kwark, M. B. Ritter, and J. L. Drewniak, "Accuracy and application of physics-based circuit models for vias," *in Proc. IMAPS 39th International Symposium on Microelectronics*, San Diego, California, USA, October 8-12, 2006.

[96] G. Selli, C. Schuster, Y. H. Kwark, M. B. Ritter, and J. L. Drewniak, "Developing a "physical" model for vias - part II: coupled and ground return vias," in *Proc. IEC DesignCon*, Santa Clara, California, USA, 2007.

[97] A. G. Williamson, "Radial-line/coaxial-line junctions: analysis and equivalent circuits," *International Journal of Electronics*, vol. 58, no. 1, pp. 91-104, January 1985.

[98] Q. Gu, M. A. Tassoudji, S. Y. Poh, R.T. Shin, and J. A. Kong, "Coupled noise analysis for adjacent vias in multilayered digital circuits," *IEEE Transactions on Circuits and Systems I: Fundamental Theory and Applications*, vol. 41, no. 12, pp. 796-804, December 1994.

[99] R. Abhari, G. V. Eleftheriades, and E. van Deventer-Perkins, "Physics-based CAD models for the analysis of vias in parallel-plate environments," *IEEE Transactions on Microwave Theory and Techniques*, vol. 49, no. 10, pp. 1697-1707, October 2001.

[100] M. Xu, Y. Ji, T. H. Hubing, T. Van Doren, and J. L. Drewniak, "Development of a closed-form expression for the input impedance of power-ground plane structures," *in Proc. IEEE International Symposium on Electromagnetic Compatibility*, Washington D.C., USA, August 21-25, 2000, pp. 77-82.

[101] Q. Li, L. Tsang, H. Chen, "Quasi-static parameters, low-frequency solutions, and full-wave solutions of a single-layered via, *Microwave and Optical Technology Letters*, vol. 35, no. 1, pp, 34 40, October 2002.

[102] J. C. Parker, "Via coupling within parallel rectangular planes," *IEEE Transactions on Electromagnetic Compatibility*, vol. 39, no. 1, pp. 17-23, February 1997.

[103] N. Na, J. Jinseong, S. Chun, M. Swaminathan, and J. Srinivasan, "Modeling and transient simulation of planes in electronic packages," *IEEE Transactions on Advanced Packaging*, vol. 23, no. 3, pp. 340-352, August 2000.

[104] G. Antonini, "A low-frequency accurate cavity model for transient analysis of power-ground structures," *IEEE Transactions on Electromagnetic Compatibility*, vol. 50, no. 1, pp. 138-148, February 2008.

[105] Z. L. Wang, O. Wada, Y. Toyota, and R. Koga, "Convergence acceleration and accuracy improvement in power bus impedance calculation with a fast algorithm using cavity modes," *IEEE Transactions on Electromagnetic Compatibility*, vol. 47, no. 1, pp. 2-8, February 2005.

[106] J. Trinkle and A. Cantoni, "Single summation expression for the rectangular power ground plane cavity," *in Proc. 16th Int. Zurich Symposium on Electromagnetic Compatibility*, Zurich, Switzerland, February 16-20, 2005, pp. 247-250.

[107] M. Hampe, V. Palanisamy, and S. Dickmann, "Single summation expression for the impedance of rectangular PCB power-bus structures loaded with multiple lumped elements," *IEEE Transactions on Electromagnetic Compatibility*, vol. 49, no. 1, pp. 58-67, February 2007.

[108] J. Trinkle and A. Cantoni, "Impedance expressions for unloaded and loaded power ground planes," *IEEE Transactions on Electromagnetic Compatibility*, vol. 50, no. 2, pp. 390-398, May 2008.

[109] A. Benalla and K. C. Gupta, "Faster computation of Z/matrices for rectangular segments in planar microstrip circuits," *IEEE Transactions on Microwave Theory and Techniques*, vol. 34, pp. 733-736, June 1986.

[110] A. R. Chada, Y. Zhang, G. Feng, J. L. Drewniak, and J. Fan, "Impedance of an infinitely large parallel-plane pair and its applications in engineering modeling," *in Proc. IEEE International Symposium on Electromagnetic Compatibility*, Austin, Texas, USA, August 17-21, 2009, pp. 78-82.

[111] R. Chadha and K. C. Gupta, "Green's functions for triangular segments in planar microwave circuits," *IEEE Transactions on microwave theory and techniques*, vol. 28, no. 10, pp. 1139-1143, October 1980.

[112] R. Chadha and K. C. Gupta, "Green's functions for circular sectors, annular rings, and annular sectors in planar microwave circuits," *IEEE Transactions on Microwave Theory and Techniques*, vol. 29, no. 1, pp.68-71, January 1981.

[113] S. H. Lee, A. Benalla, and K. C. Gupta, "Faster computation of Z-matrices for triangular segments in planar circuits," *International Journal of Microwave and Millimeter-Wave Computed-Aided Engineering*, vol. 2, no. 2, pp. 98-107, 1992.

[114] Z. L. Wang, O. Wada, Y. Toyota, and R. Koga, "Application of segmentation method to analysis of power/ground plane resonance in multilayer PCBs," *in Proc. IEEE 3rd International Symposium on Electromagnetic Compatibility*, Beijing, China, May 21-24, 2002, pp. 775-778.

[115] Z. L. Wang, O. Wada, Y. Toyota, and R. Koga, "Analysis of resonance characteristics of a power bus with rectangle and triangle elements in multilayer PCBs," *in Proc. Asia-Pacific Conference on Environmental Electromagnetics CEEM*, November 4-7, 2003, pp.73-76.

[116] Z. L. Wang, O. Wada, Y. Toyota, and R. Koga, "Modeling of gapped power bus structures for isolation using cavity modes and segmentation," *IEEE Transactions on Electromagnetic Compatibility*, vol. 47, no. 2, pp. 210-218, May 2005.

[117] B. Osterhoff, "Accuracy and efficiency analysis of techniques to compute the impedance for irregular power planes," *Studienarbeit, Institut für Theoretische Elektrotechnik, Technische Universität Hamburg-Harburg*, January 2010.

[118] Y. Zhang, J. Fan, G. Selli, M. Cocchini, and D. P. Francesco, "Analytical evaluation of via-plate capacitance for multilayer printed circuit boards and packages," *IEEE Transactions on Microwave Theory and Techniques*, vol. 56, no. 9, pp. 2118–2128, September 2008.

[119] Ansoft Corporation, Q3D Extractor, Ver. 8 [Online], Pittsburgh, USA. Information available: http://www.ansoft.com (March. 2010).

[120] CST Corporation, CST EM Studio Ver. 2009 [Online], Darmstadt, Germany. Information available: http://www.cst.com (March 2010).

[121] C. R. Paul, "Decoupling the multiconductor transmission line equations," *IEEE Transactions on Microwave Theory and Techniques*, vol. 44, no. 8, pp. 1429-1440, August 1996.

[122] C. R. Paul, "Analysis of multiconductor transmission lines," New York, USA: *Wiley Series in Microwave and Optical Engineering*, 1994.

[123] S. Müller, "Including multiconductor transmission lines in a quasi-analytical model for multilayer structures," *Diplomarbeit. Institut für Theoretische Elektrotechnik, Technische Universität Hamburg-Harburg*, November 2009.

[124] H. Marko, "Theorie linearer Zweipole, Vierpole und Mehrtore," Stuttgart, Germany: *S. Hirzel Verlag,* Stuttgart, 1971.

[125] R. Chadha and K. C. Gupta, "Segmentation method using impedance matrices for analysis of planar microwave circuits," *IEEE Transactions on Microwave Theory and Techniques*, vol. 29, no. 1, pp. 71-74, January 1981.

[126] Institut für Theoretische Elektrotechnik, TUHH, CONCEPT-II [Online], Hamburg, Germany. Information available: http://www.tet.tu-harburg.de (March 2010).

[127] Ansoft Corporation, High Frequency Structure Simulator (HFSS) Ver. 11 [Online], Pittsburgh, USA. Information available: http://www.ansoft.com (March. 2010).

[128] Y. H. Kwark, C. Schuster, L. Shan, C. Baks, and J. Trewhella, "The recessed probe launch – A new signal launch for high frequency characterization of board level packaging," *in Proc. IEC DesignCon Conference*, Santa Clara, California, USA, 2005.

[129] A. P. Duffy, A. J. Martin, A. Orlandi, G. Antonini, T. M. Benson, and M. S. Woolfson, "Feature selective validation (FSV) for validation of computational electromagnetics. Part I - The FSV method," *IEEE Transactions on Electromagnetic Compatibility*, vol. 48, no. 3, pp. 449-459, August 2006.

[130] A. Orlandi, A. P. Duffy, B. Archambeault, G. Antonini, D. E. Coleby, and S. Connor, "Feature selective validation (FSV) for validation of computational electromagnetics. Part II - Assessment of FSV performance," *IEEE Transactions on Electromagnetic Compatibility*, vol. 48, no. 3, pp. 460-467, August 2006.

[131] Standard for validation of computational electromagnetics computer modeling and simulation, *IEEE P1597.1*, 2008.

[132] University of L'Aquila EMC Laboratory, FSV routine [Online], L'Aquila, Italy. Available at: http://ing.univaq.it/uaqemc/FSV_4_0_3L (Feb. 2009)

[133] D. E. Bockelman and W. R. Eisenstadt, "Pure-mode network analyzer for on-wafer measurements of mixed-mode S-parameters of differential circuits," *IEEE Transactions on Microwave Theory and Techniques*, vol. 45, no. 7, pp, 1071-1077, July 1997.

[134] W. Fan, A. C. W. Lu, L. L. Wai, and B. K. Lok, "Mixed-mode S-parameter characterization of differential structures," *in Proc. 5th Conference in Electronics Packaging Technology EPTC*, Singapore, December 10-12, 2003, pp. 533-537.

[135] Park Electrochemical Corp. "Nelco advanced circuitry materials," [Online], California, USA. Information available: http://www.parkelectro.com/parkelectro/images/n4000-13.pdf (March 2010).

[136] X. Gu, B. Wu, C. Baks, and L. Tsang, "Fast full wave analysis of PCB via arrays with model-to-hardware correlation," *in Proc. IEEE 18th Conference on Electric Performance of Electronic Packages and Systems EPEPS*, Portland, Oregon, USA, October 19-21, 2009, pp. 175-178.

[137] Mathworks Corporation, Matlab Ver. 2009 [Online], Natick, MA, USA. Information available: http://www.mathworks.com (March 2010).

[138] P. J. Acklam, "MATLAB array manipulation tips and tricks," [Online], Information available http://home.online.no/~pjacklam (March 2010).

[139] D. Timmermann, "Implementation of efficient algorithms for parallel-plate impedance calculation," *Studienarbeit, Institut für Theoretische Elektrotechnik, Technische Universität Hamburg-Harburg*, October 2009.

[140] S. Chapman, "Fortran 95/2003 for scientists & engineers," 3rd Ed., New York, USA: *McGraw-Hill*, 2008.

[141] N. H. Malik, "A review of the charge simulation method and its applications," *IEEE Transactions on Electrical Insulation*, vol 24, no. 1, pp. 3-20, February 1989.

[142] S. Maeda, T. Kashiwa, and I. Fukai, "Full wave analysis of propagation characteristic of a through hole using the finite-difference time domain method," *IEEE Transactions on Microwave Theory and Techniques*, vol. 39, no. 12, pp, 2154-2159, December 1991.

[143] U. L. Pfeiffer, "Low-loss contact pad with tuned impedance for operation at millimeter wave frequencies," *in Proc. IEEE Workshop on Signal Propagation on Interconnects SPI*, Garmisch-Partenkirchen, Germany, May 10-13, 2005, pp. 61-64.

[144] B. Gaucher, Y. Kwark, C. Schuster, "High performance resonant element", *United States Patent Application US 2007/0286293 A1*, December 2007.

[145] M. Mondal, B. Mutnury, P. Patel, S. Connor, B. Archambeault, and M. Cases, "Electrical analysis of multi-board PCB systems with differential signaling considering non-ideal common ground connection," *in Proc. Electrical Performance of Electronic Packaging Conference EPEP*, Atlanta, Georgia, USA, October 29-31, 2007, pp. 37 - 40.

[146] B. Archambeault, J. C. Diepenbrock, and S. Connor, "EMI emissions from mismatches in high speed differential signal traces and cables," *in Proc. IEEE Int. Symposium on Electromagnetic Compatibility*, Honolulu, Hawaii, July 8-13, 2007, pp. 1-6.

[147] J. L. Knighten, B. Archambeault, J. Fan, G. Selli, A. Rajagopal, S. Connor, and J. L. Drewniak, "PDN design strategies: IV. Sources of PDN noise," *IEEE EMC Society Newsletter*, issue no. 212, pp. 54-64, Winter 2007.

[148] J. Fan, M. Cocchini, B. Archambeault, J. L. Knighten, J. L. Drewniak, and S. Connor, "Noise coupling between signal and power/ground nets due to signal vias transitioning through power/ground plane pair," *IEEE Symposium on Electromagnetic Compatibility*, Detroit, USA, August 18-22, 2008.

[149] A. Ferrero and M. Pirola, "Generalized mixed-mode S-parameters," *IEEE Transactions on Microwave Theory and Techniques*, vol. 54, no. 1, pp, 458-463, January 2006.

[150] S. Van den Berghe, F. Olyslager, D. De Zutter, J. De Moerloose, and W. Temmerman, "Study of the ground bounce caused by power planes resonance," *IEEE Transactions Electromagnetic Compatibility*, vol. 40, no. 2, pp. 111-119, May 1998.

[151] D. A. Frickey, "Conversions between S, Z, Y, h, ABCD, and T parameters which are valid for complex source and load impedance," *IEEE Transactions on Microwave Theory and Techniques*, vol. 42, no. 2, pp. 205-211, February 1994.

[152] M. Stumpf, O. Kröning, and M. Leone, "Power-bus modeling using 2D-integral-equation formulation," *in Proc. 20th International Zurich Symposium on Electromagnetic Compatibility*, Zurich, Switzerland, January 12-16, 2009, pp.189-192.

[153] C. A. Balanis, "Antenna theory," 2nd Ed., New York, USA: *Wiley*, 1997.

[154] M. Leone, "The radiation of a rectangular power-bus structure at multiple cavity-mode resonances," *IEEE Trans. Electromagnetic Compatibility*, vol. 45, no. 3, pp. 486-492, August 2003.

[155] D. M. Pozar, "Microwave engineering," 3rd Ed., New York, USA: *John Wiley & Sons*, 2005.

[156] N. Marcuvitz, [C. G. Montgomery, R. H. Dicke, and E. M. Purcell, Ed.], "Radial transmission lines," *in Principles of Microwave Circuits*, rev Ed. vol. 25, London, United Kingdom: IEEE, 1987.

[157] N. Marcuvitz, "Waveguide handbook," in IEE Electromagnetic Waves Series 21, London, United Kingdom: *Peter Peregrinus Ltd.*, 1986.

[158] EIA/IBIS Open Forum, "Touchstone® (.SnP) file format specification. Rev 1," [Online]. Information available: http://vhdl.org/ibis/connector/ touchstone_spec11.pdf (March 2010).

[159] R. Rimolo-Donadio, "VPF-Tool documentation," Internal Document. *Institut für Theoretische Elektrotechnik, Technische Universität Harburg-Harburg*, 2010.

Curriculum Vitae

Renato Rimolo Donadio

Born on March 25, 1977,

in San José, Costa Rica,

Central America.

University Studies

11/2006 – 12/2010

Technische Universität Hamburg-Harburg, Hamburg (TUHH), Germany.

Doctoral studies on "Development, Validation and Application of Semi-Analytical Interconnect Models for Efficient Simulation of Multilayer Substrates".

10/2004 – 10/2006

Technische Universität Hamburg-Harburg, Hamburg (TUHH), Germany.

Master of Science in Microelectronics and Microsystems. Graduated with distinction.

Thesis on "Design of a Configurable Fully-Differential Operational Transconductance Amplifier for Low Power Sigma-Delta Modulators".

02/2003 – 06/2004

Technical University of Costa Rica (ITCR), Cartago, Costa Rica.

Licentiate in Electronic Engineering. Graduated with honors.

02/1994 – 12/1998

Technical University of Costa Rica (ITCR), Cartago, Costa Rica.

Bachelor in Electronic Engineering. Graduated with distinction.

Professional and Academic Experience

Since 11/2006

Technische Universität Hamburg-Harburg (TUHH), Hamburg, Germany.

Research Assistant at the Institute of Electromagnetic Theory (TET).

07/2007 – 09/2007

IBM T. J. Watson Research Center, Yorktown Heights, New York, USA.

Intern at the High-Speed I/O Subsystems and Packaging Group.

09/2005 – 12/2005

CNM-IMSE Institute of Microelectronics, Seville, Spain.

Intern at the Digital Design Group.

07/2003 – 09/2004

Technical University of Costa Rica (ITCR), Cartago, Costa Rica.

Instructor at the Electronic Engineering Department.

05/2001 – 08/2003

Latin University of Costa Rica (ULCR), San Pedro, Costa Rica.

Instructor at the Computer Science Faculty.

06/1999 – 12/2002

BTC of Costa Rica – CELLE S.A., San Pedro, Costa Rica.

IT Development and Support Engineer.

Hamburg, 17.12.2010